Microencapsulation - Processes, Technologies and Industrial Applications

Edited by Fabien Salaün

Published in London, United Kingdom

IntechOpen

Supporting open minds since 2005

Microencapsulation - Processes, Technologies and Industrial Applications
http://dx.doi.org/10.5772/intechopen.73747
Edited by Fabien Salaün

Contributors
Tri Yuni Yuni Hendrawati, Alvika Meta Sari, Ratri Ariatmi Nugrahani, Muhamad Iqbal Syauqi Rahman, Mariem Ben Jemaa, Hanen Falleh, Riadh Ksouri, Ekasit Onsaard, Wiriya Onsaard, Leilane Costa De Conto, Pedro Henrique Rodrigues Do Amaral, Patrícia Lopes Andrade, Mahadevappa Kariduraganavar, Shaluah Vijeth, Geetha Heggannavar, Seng-Neon Gan, Nurshafiza Shahabudin, Chigozie Uzoh, Raphael Stone Odera

First published in London, United Kingdom, 2019 by IntechOpen
IntechOpen is the global imprint of INTECHOPEN LIMITED, registered in England and Wales, registration number: 11086078, The Shard, 25th floor, 32 London Bridge Street
London, SE19SG – United Kingdom
Printed in Croatia

British Library Cataloguing-in-Publication Data
A catalogue record for this book is available from the British Library

Additional hard and PDF copies can be obtained from orders@intechopen.com

Microencapsulation - Processes, Technologies and Industrial Applications
Edited by Fabien Salaün
p. cm.
Print ISBN 978-1-83881-869-2
Online ISBN 978-1-83881-870-8
eBook (PDF) ISBN 978-1-83881-871-5

For EU product safety concerns:
IN TECH d.o.o., Prolaz Marije Krucifikse Kozulić 3, 51000 Rijeka, Croatia, info@intechopen.com or visit our website at intechopen.com.

Meet the editor

Fabien Salaün received his Master's degree in Organic and Macromolecular Chemistry at Lille 1 University (France) in 2001. In 2004, he received his PhD in Organic and Macromolecular Chemistry. In 2007, he obtained the position of Assistant Professor at ENSAIT (France) and from September 2017 he became full-time Professor in Polymers Science and Textile Technologies. He has coauthored more than 45 peer-reviewed papers (885 total citations; h Index: 17), 14 book chapters, and several communications to national and international congresses. His main research activities include the formulation of microencapsulation processes, the synthesis of nano- and microcapsules, the surface functionalization of textiles, the modification of polymers and their physicochemical and thermomechanical characterization, and the investigation of structure–property relationships.

Contents

Preface

Introduction to microencapsulation

The term microencapsulation comes from the Greek "mikros," which means small, and from the Latin "en" and "capsula," which mean respectively in and a small box. This technology refers to the formation of polymer particles with diameters ranging from nanometers to millimeters, containing an active substance regardless of its physical state, i.e., in solid, liquid, or gaseous form, and which may have several types of morphologies. Over the past two decades, microencapsulation has been a growing field with applications in many technological disciplines. However, the principle of encapsulation is not new; biochemistry is one of the great founding principles of life, and the presence of a membrane allows the containment of essential molecules in cells. Nature is full of examples of encapsulation, from the macro- to the nanoscale, to protect material from the surrounding environment, such as a seed in a coat, a bird's egg, or a cell in a membrane. Thus, the development of microencapsulation processes is similar to an imitation of nature, to design innovative systems to immobilize, structure, or release the active ingredient. Since encapsulation has applications in various industrial fields, many definitions depend on the needs in a specific field. The most general definition probably corresponds to the trapping of a compound or system in a dispersed material for its immobilization, protection, transfer control, structuring, and functionalization.

The development of the field of microencapsulation is closely linked to the progress of many advanced technologies to produce an innovative component for a specific application or industrial field, and thus to create a new application or market for products. One of the benefits of microencapsulation is that it combines various technologies from other unrelated industrial fields, which are combined to develop a new product. Also, in recent decades, the development of microencapsulation technology has also been accompanied by the reduction of particle size to the manometric scale, which is used in some advanced markets.

The first mechanical microencapsulation technologies were developed at the end of the nineteenth century, and more particularly from spray coating and drying. It was only in the 1930s that the first attempts at phase coacervation microencapsulation were made by Bungenburg de Jong and Kass. A few years later, B. Green, a research chemist at the National Cash Register Company in Dayton, became interested in this technology and studied how the concept of microencapsulation could have a potential application in document copying. From the concept of phase coacervation developed by Bungenburg de Jong and Kass, he prepared the first gelatin-based microcapsules. The development of this technology from the laboratory to industrial scale took several years, and it was not until 1951 that NCR introduced the first carbonless copy paper. Green filed a patent for microencapsulation on June 30, 1953.

At the end of the 1950s, microencapsulation was gradually introduced in the pharmaceutical and chemical sectors. Since then, microencapsulation processes and technologies have been continuously improved, modified, and adapted for various purposes and/or uses. Thus, the development of new methods and the expansion of the range of new polymer materials adapted to the different trapping techniques have made it possible to create new products and applications according to customer supply and demand or sometimes as a result of merged projects. The development of microencapsulation technology has been characterized in recent decades by rapid growth in patent applications and scientific articles, reflecting the interest of industrial and academic research in this subject. Thus, this technology has a host of potential applications in a wide range of industrial sectors, i.e., cosmetics, pharmaceutical and medical, electronic, waste treatment, printing, food, agriculture, biotechnology, chemical, textile, etc.

In the coming years, the size of the global microencapsulation market is expected to reach $19.34 billion by 2025, with a compound annual growth rate of 13.6%. This increase is mainly driven by the food and beverage industry, where the demand for microencapsulated flavors, probiotic bacteria, and immobilized cells or enzymes will increase further. The microencapsulation markets are mainly dependent on geographical regions. In North America, for example, this technology has mainly been introduced in the pharmaceutical, food, and personal care industries, while the textile industry could also play a more critical role in the future. In Western Europe, the second largest regional market in 2017, the market is overgrowing regarding the availability of raw materials, coating technologies, and fields of application, and more specifically in the pharmaceutical and cosmetics industries. The Asia-Pacific market is expected to be one of the main growth areas, driven in particular by the rapid growth of the pharmaceutical and food industries and the growth of the detergent market.

Microencapsulation concept

Microencapsulation is a process by which individual elements of an active substance are stored within a shell, surrounded or coated with a continuous film of polymeric or inorganic material to produce particles in the micrometer to millimeter range, for protection and/or controlled release. The particles obtained by this process are called microparticles, microcapsules, and microspheres according to their morphologies and internal structure. For particles with a size range below 1 μm, the terms nanoparticles, nanocapsules, and nanospheres are often used. When average diameters higher than 1000 μm are obtained, the term "macrocapsules" is adequate. The nomenclature used to define the different elements of the encapsulated product includes terms for the shell, i.e., wall, coating, membrane material, and for the core material, i.e., active agent, payload, or internal phase. Various compounds from different origins such as dyes, proteins, fragrances, monomers, catalysts, etc. can be encapsulated with different types of wall material, including natural polymer (gelatin, cellulose, chitosan, etc.), artificial polymer (cellulosic derivatives, etc.), and synthetic polymers (polyamide, polyester, etc.), for a loading content between 5 and 90% of the microparticles in weight.

Microcapsules can exhibit a wide range of geometries and structures. The morphologies, geometries, or structures of the microcapsules depend mainly on the physicochemical characteristics of the core material and the process used to induce membrane formation. Thus, microparticles may have regular or irregular shapes, and their morphology may be described as mononuclear or core/shell structure, multinuclear or polynuclear particles, and matrix particle or microsphere. Microspheres consist of a polymeric network structure in which tiny particles of an active substance are distributed homogeneously, whereas microcapsules or core/shell structures exhibit a reservoir structure, i.e., the core substance is surrounded by a distinct layer. Furthermore, the layer and the core numbers are not limited to a single one; double-layered microcapsule shells, multilayer microcapsules, or dual-core particles are found in the literature, and the microparticles can also be obtained from clusters of microcapsules. The process choice and the formulation parameters, such as the polymer/solvent system, dictate their characteristics as well as their functional properties.

Benefits of microencapsulation

According to the end-use, various characteristics of the microcapsules may be desired to design the final product. The size and shape of the particles, physicochemical properties of the shell, compatibility, and permeability are some of the main characteristics to consider in the choice the processes, taking in account the physicochemical properties of the active substance. The porosity of the shell is also controlled to complete the final application.

Protection and shelf life enhancement

Some of the active substances used are chemically fragile, volatile, or unstable, and cannot be directly used without being entrapped in a capsule. One of the main advantages of using microencapsulation technology is its ability not only to protect active substances from the surrounding environment thereby increasing the shelf life of the product and its activity especially for fragrance or cosmetic applications, but also to prevent interaction with other compounds in the system or other components. The microcapsule shell capsules can prevent the evaporation of the volatile substance and protect both workers and end-users from exposure to toxic or hazardous substances; therefore, the active substances are handled more safely under this form before processing. They also transform a soluble compound in a temporarily insoluble form or realize the complex mixture of various components in a single capsule for a specific application. This process also allows masking an odor or an unpleasant fragrance of the active compounds during manufacture and end-use.

Controlled release

The use of microencapsulation technology for a controlled release application is one of the best ways to increase efficiency and minimize environmental damage. It can achieve a controlled delay of the release of the active substance until a stimulus is encountered at a specific rate, time, or situation, i.e., heating, moisture, mechanical pressure, etc. In this case, the shell acts as a protective barrier to prevent diffusion and mass loss of the active substance, and therefore maintain intact the core compounds until final use.

Compatibility

The transformation of an active liquid into a pseudo-solid or powdered product can not only limit agglutination but also improve the mixing of incompatible compounds. Moreover, if one considers, for example, a textile application, the effectiveness of a binder in fixing microcapsules on a surface depends on the compatibility of the different interfaces between the elements involved in the process, and is closely related to the individual nature and chemical structure of each component.

Microencapsulation technologies

More than 200 microencapsulation methods are described in the scientific literature and patents, and most of them include three necessary steps, namely containment of the central component, formation of microparticles, and hardening of the envelope. These methods are generally divided or classified into three main groups, which are based on the mechanisms governing membrane formation, namely mechanical, chemical, and physicochemical processes. The choice of one method over another is often dictated by the cost of treatment, the use or not of organic solvents, and the consideration of health and environmental aspects. The interactions between polymers and solvents in the microencapsulation process probably have the most critical effect on the morphology and properties of the particles obtained. Thus, each encapsulation step is affected by the solvency of the oil phase, and therefore to form a separate membrane or shell, the solvent used must promote the precipitation of the polymer in the early reaction stage, and also allow the continuous diffusion of monomers through the existing membrane to allow its growth.

The most popular methods include interfacial, in-situ, and suspension polymerization methods for chemical processes, simple or complex phase coacervation for physicochemical processes, and spray drying for mechanical processes. Whatever the process selected, it includes two main steps, e.g., the emulsification step, which determines the size and size distribution of the microcapsules, and the formation of the capsules. The first step is affected immediately by physical parameters such as apparatus configuration, stirring rate, and volume ratio of the two phases, and by physicochemical properties such as interfacial tension, viscosities, densities, and chemical compositions of the two phases used. The formation of microcapsules is also related to the use of surfactants, which influences not only the mean diameter but also the stability of the dispersion. The surfactants or the colloid have two leading roles, i.e., to reduce the interfacial tension between oil and aqueous phases to allow the formation of smaller microcapsules, and to limit or prevent coalescence by its adsorption on the oil/water interface by forming a layer around the dispersed droplets. Shell formation is mainly governed by kinetic factors, i.e., the ability of the monomers, pre-polymer, or polymer to react or to cross-link, and thermodynamic factors, i.e., the minimum total free energy exchange in the system. Furthermore, the choice of the polymer system for shell synthesis needs to be considered regarding the application and availability of the material.

Outline of the book

This book is intended to provide an overview and review of the latest developments in microencapsulation processes and technologies for various fields of applications. The general theme and purpose are to provide the reader with a current and general overview of the existing microencapsulation systems and to emphasize various methods of preparation, characterization, evaluation, and potential applications in various fields such as medicine, food, agricultural, and composites. It targets readers, including researchers in materials science processing and/or formulation and microencapsulation science, engineers in the area of microcapsule development, and students in colleges and universities.

The book has been contributed to by a panel of international researchers and experts in the field of microencapsulation and covers various aspects of the research and development of the processes, technologies, and applications. It is composed of eight chapters, which can be divided into three main parts. The preface provides background information on microencapsulation technology, including a brief overview of this theme, and the book's structure. Thus, the chapters are arranged logically according to the materials used, the methods of preparation, and applications of microcapsules in certain fields.

The first part involves the classification of the shell and core materials used in the main processes. Chapter 1 by S. Vijeth, G.B. Heggannavar, and M.Y. Kariduraganavar is concerned with biodegradable and responsive polymers as wall materials for developing stable and safe micro/nanocapsules, and also covers the development and progress made in the selection of suitable polymeric shells for drug-delivery systems. Chapter 2 by E. Onsaard and W. Onsaard deals with the use of vegetable power as a core component and focuses on emulsion preparation using multilayer emulsions followed by the spray-drying technique to obtain a vegetable oil powder.

The second part of the book, which mainly focuses on nanoemulsions and the spray-drying method, starts with a chapter by M.B. Jemaa, H. Falleh, and R. Ksouri on the concept of a nanoemulsion-based delivery system to immobilize active compounds such as essential oils. The following chapter by T.Y. Hendrawati and A.M. Sari presents a detailed explanation of the preparation and characterization of herbal compounds-loaded microcapsules. The authors clearly define the guidelines for choosing the appropriate working conditions to encapsulate turmeric by the spray-drying method, and then focus on the encapsulation of aloe vera.

In the third part of the book, examples of applications are presented. S.N. Gan and N. Shahabudin discuss the methods of preparation of microcapsules for self-healing materials. In his comprehensive and exhaustive chapter, P.H.R. do Amaral underlines the potential uses of microcapsules in the food industry and emphasizes recent progress made in the last decade in understanding production methods. In the last chapter, C.F. Uzoh and O.D. Onukwuli propose the application of capsules in the agricultural field to control the release of fertilizer as a function of coating thickness.

This book will provide scientists, researchers, and students in the field of microencapsulation formulation, processes, technologies, material science, and bioengineering, and professionals in the pharmaceutical, biotechnology, food, cosmetics, textiles, and agricultural industries with an essential compendium of concepts and practical cases for their daily activities. I want to thank all the contributors of this book. I also wish to express my heartfelt gratitude to the team at IntechOpen for their help with this project.

Fabien Salaün
Professor
ENSAIT, GEMTEX – Laboratoire de Génie et Matériaux Textiles,
Lille, France

Section 1

Core and Shell Materials

Chapter 1

Encapsulating Wall Materials for Micro-/Nanocapsules

Shaluah Vijeth, Geetha B. Heggannavar, Mahadevappa Y. Kariduraganavar

Abstract

Wall materials play a vital role in the development of micro-/nanocapsules to protect the bioactive compounds against external factors. The encapsulation process and the type of polymers exert a direct impact on the development of bioactive micro-/nanocapsules, which greatly reflect in encapsulation efficiency, solubility, stability, surface permeability, and release profile of desired bioactive compounds. Among the polymers, biodegradable polymeric materials have been the focus for various applications in food, pharmaceutical, and cosmetic industries. Thus, this chapter focuses on different encapsulation techniques and the importance of biodegradable polymers employed as wall materials for developing stable and safe micro-/nanocapsules. Among the natural polymers, protein- and polysaccharide-based polymers are widely used. Similarly, the most commonly used synthetic polymers are polycaprolactone, poly(lactic-co-glycolic acid), and polyethylene glycol. Synthetic polymers have been classified based on their exogenous and endogenous responsive natures. At the end, we have also discussed on the applications of biodegradable polymers employed in the development of micro-/nanocapsules. To compile this chapter and to provide adequate information to the readers, we have explored various sources, such as reviews, research articles, books, and book chapters including Google sites.

Keywords: biodegradable polymers, microcapsules, nanocapsules, responsive polymers

1. Introduction

Encapsulation is a process in which tiny particles or droplets are surrounded by a coating to form capsules. Microcapsules and nanocapsules are small spheres with a uniform wall around it. The material inside the capsule is referred as the core, internal phase, or fill, whereas the wall is called as the shell, coating, or membrane. The core material may be liquid or solid, active constituents, stabilizers, diluents, excipients, and release-rate retardants or accelerators.

Encapsulation can be done for multiple reasons. The primary purpose of encapsulation is either for sustained or prolonged drug release. For orally delivered drugs, this method has been widely used for masking taste and odor to improve patient compliance, reduce toxicity, and gastrointestinal irritation. This method can be used to convert liquid drugs to a free flowing powder form, prevent vaporization of volatile drugs, alter the site of absorption, and to prevent incompatibility among the drugs. The drugs which are sensitive to oxygen, moisture, light, or pH changes can be stabilized by encapsulation. Depending on the applications, the wall material of the capsules is designed to serve the desired specific purpose.

Thus, the wall material is the most vital component in any capsule. The selection of appropriate wall material decides the physical and chemical properties of the resultant micro-/nanocapsules. The wall material should have the properties like being inert toward core active ingredients, stabilize the core material, film-forming, pliable and tasteless, non-hygroscopic, moderate viscosity, economical and soluble in an aqueous media or solvent, or melting, the coating may be flexible, brittle, hard, and thin, controlled release at the specific site under specific conditions.

Understanding the importance of polymers as wall materials, we have attempted to give a review of wall materials used in micro- and nanocapsules for the sustained/controlled delivery of drugs. A detailed account on encapsulation techniques is given. The biodegradable materials employed as wall materials are discussed adequately. The advantages of biodegradable materials including their limitations are covered in the chapter. The release profiles were discussed based on both exogenous and endogenous responsiveness of the wall materials. At the end of the chapter, future prospects and challenges of the wall materials are highlighted.

2. Encapsulation techniques

Microencapsulation of active compounds can be achieved by physical and chemical methods. Though these techniques are neither purely physical nor purely chemical, they are classified as physical and chemical methods based on the predominant or primary principle involved. In **Table 1**, we have listed commonly used physical and chemical methods used for encapsulation.

Physical methods	Chemical methods
Air suspension coating	Solvent evaporation
Coacervation	Polymerization
Centrifugal extrusion	
Pan coating	
Spray drying	
Fluidized bed technology	

Table 1.
Encapsulation techniques.

3. Polymers as wall materials

3.1 Natural polymers

Natural polymers are broadly classified into protein-based polymers and polysaccharide-based polymers. Albumin and gelatin are the examples of protein-based polymers. Polysaccharide-based polymers are agarose, alginate, hyaluronic acid, dextran, chitosan, etc. These natural polymers are highly biodegradable and biocompatible in nature.

3.1.1 Protein-based polymers

3.1.1.1 Albumin

Albumin is a biodegradable and water soluble protein and thus plays an important role in the circulating system. It is involved in osmotic pressure regulation, binding, and transport of nutrients to the cells that can be obtained from a variety of sources including egg white, bovine serum, and human serum. It is stable in the pH range of 4–9 and can be heated at 60°C up to 10 h without any deleterious effects [1]. It undergoes degradation by protease enzymes, which helps the microcapsules to release the drugs in the small intestine. It also facilitates the release of therapeutic cargo from nanocapsules inside the endosomes.

3.1.1.2 Gelatin

Gelatin is biodegradable, inexpensive, easily sterilized, non-pyrogenic, nontoxic, non-immunogenic, and easy to be crosslinked or modified chemically. Gelatin has many ionizable groups, such as carboxyl, amino, phenol, guanidine, and imidazole, which are potential sites for conjugation or chemical modifications. Chemical crosslinking agents like glutaraldehyde improves the integrity and performance of the gelatin and provides gelatin with greater stability, shape, and increased circulation time *in vivo* [2]. The degree of crosslinking determines the release of drugs from the gelatin capsules. Thus, gelatin is regarded as a safe excipient approved by the US FDA for pharmaceutical applications.

3.1.2 Polysaccharide-based polymers

3.1.2.1 Chitosan

Chitosan, the second most abundant polysaccharide in nature, is a promising biopolymer widely used in biomedical and pharmaceutical fields like wound dressing, tissue engineering and drug delivery. It is produced from chitin which is the structural element found in the exoskeleton of crustaceans like shrimps, lobsters, and crabs. Chitosan has been reported to exhibit many therapeutic properties, such as activation of immune response, cholesterol lowering activity, anti-hypertensive activity, inhibition of growth of microorganism, pain alleviation, and promotion of hemostasis and epidermal cell growth [3]. This is all due to the favorable pharmaceutical properties of chitosan, such as biocompatibility, low production cost, ability to bind some organic compounds, susceptibility to enzymatic hydrolysis, and nontoxicity.

Chitosan is considered as the most important polysaccharide-based polymer owing to its cationic character based on its primary amino groups, which are responsible for its versatile properties, such as mucoadhesion (improves pulmonary drug delivery), controlled drug release, transfection, in situ gelation, and efflux pump inhibitory properties and permeation enhancement [4]. The major drawback is its poor solubility at physiological pH due to partial protonation of the amino groups in the presence of proteolytic enzymes and thereby causing presystemic metabolism of drugs in intestinal and gastric fluids. To overcome these inherent drawbacks, various derivatives of chitosan, such as carboxylated, different conjugates, thiolated and acylated chitosan have been used in drug-delivery systems [3].

Chitosan is produced by the deacetylation of chitin. The degree of deacetylation is related to chitosan's crystallinity and degradation rates. Chitosan's solubility can also be improved when the primary amino group is protonated at low pH. The viscosity of chitosan solution increases with increasing the concentration of chitosan [5]. These properties and the ease with which it can be modified makes chitosan a versatile and bioactive polymer for its use in encapsulation.

3.1.2.2 Hyaluronic acid

Hyaluronic acid (HA) is a nonsulfated glycosaminoglycan, comprising a relatively simple linear structure of alternating units of D-glucuronic acid and N-acetyl-D-glucosamine, linked via β-1,3- and β-1,4-glycosidic bonds. Hyaluronic acid is biodegradable, biocompatible, nontoxic, and non-immunogenic glycosaminoglycan distributed widely in connective, epithelial, and neural tissues.

The cluster of differentiation (CD) protein CD44 is the main HA binding receptor. CD44 is involved in the interaction between HA and the surface of specific cells, in cell proliferation, in cellular adhesion processes (aggregation and migration), angiogenesis, in cell survival and endocytosis of HA. CD44 receptor is also overexpressed in many types of tumors and this overexpression is related to tumor invasion and tumor metastasis, which makes HA a promising candidate for intracellular delivery of imaging and anticancer agents exploiting a receptor-mediated active targeting strategy. HA also interacts with hyaluronan receptor for endocytosis (HARE), lymphatic vessel endothelium receptor-1 (LYVE-1), and intracellular adhesion molecule-1 (ICAM-1), serum-derived hyaluronan-associated protein (SHAP), Brevican and Neurocan (brain and nervous tissue-specific HA and proteoglycan binding proteins), hyaluronan-binding protein 1 (HABP1) and toll-like receptors (TLRs), and all of which have specific functions. This is known as receptor-ligand interaction, which can be exploited to achieve receptor-mediated active targeting strategy [6, 7]. HA has been bioconjugated with anticancer drugs, like paclitaxel, doxorubicin, cisplatin, etc. and anti-inflammatory drugs like methotrexate, dexamethasone, methylprednisolone, etc. to achieve receptor-mediated endocytosis [8]. HA polymer has also been used in the treatment of osteoarthritis, in ocular and plastic surgery, and in tissue engineering.

3.2 Synthetic polymers

Over the past 5–6 decades, biodegradable polymers have gained tremendous attention due to their growing applications in biomaterials, drug-delivery systems, tissue engineering, and medical devices. Chemists, biologists, physicians, and engineers have collaboratively made significant advancements in these applications. The most commonly used synthetic polymers in micro-/nanocapsules for drug-delivery applications are poly(ε-caprolactone), poly(lactic-co-glycolic acid), and polyethylene glycol.

3.2.1 Poly(ε-caprolactone)

Poly(ε-caprolactone) (PCL) is a semicrystalline aliphatic polyester with glass transition temperature and melting temperature of about −60 and 60°C, respectively [9]. PCL mixes well with other polymers to form blends that impart good physical and chemical properties to achieve desired properties like swelling, porosity, and stability in different media. Microcapsulation or nanocapsulation with PCL has many advantages like modulation of drug release, control of drug penetration/permeation into the skin, and improve photochemical stability and pharmacological response. Due to its long degradation times, PCL has found many applications in tissue engineering and prolonged drug release. PCL is approved by the US Food and

Drug Administration (FDA) and has found numerous applications in implants and surgical absorbable sutures due to its biocompatibility and slow biodegradability.

3.2.2 Poly(lactic-co-glycolic acid)

Poly(lactic-co-glycolic acid) (PLGA) is the most extensively studied degradable polymer to date. PLGA is an aliphatic polyester and it undergoes hydrolysis in the body to produce biodegradable metabolite monomers such as lactic acid and glycolic acid. During metabolism in the body via the Krebs cycle, carbon dioxide, and water are removed and thereby toxicity is minimized [10]. PLGA is approved by the US FDA for use in drug-delivery systems due to its biodegradability with tissue and cells, drug biocompatibility, suitable biodegradation kinetics, mechanical properties, and ease of processing. Thus, PLGA based microcapsules and nanocapsules are the most viable candidates for drug-delivery systems, anticancer agents, bio-imaging, and vaccine immunotherapy.

3.2.3 Poly(ethylene glycol)

Polyethylene glycol (PEG) is also the most widely used "stealth" polymer in drug delivery. It is approved by the US FDA and considered to be safe. Coating of nanocapsules with PEG generates a hydration layer due to its hydrophilic nature and forms a steric barrier. This steric hindrance effect helps the nanocapsules to avoid interactions with neighboring capsules and blood components like immunogenic cells [11]. PEG coating on nanocapsules shields it from aggregation, opsonization, and phagocytosis by reticuloendothelial system. The lack of immunogenicity confers PEG-coated nanocapsules with prolonged systemic circulation time which in turn leads to enhance absorption due to enhanced permeation and retention effect. PEGylation has become a mainstay in fabrication of drug-delivery systems that require high doses of toxic drugs with prolonged duration of action.

4. Sensitive polymers

The design of polymeric drug-delivery systems has matured to exploit local biochemical changes in pathological states to trigger drug release. In a classical example, organs or tissues with cancer are characterized by a shift in homeostasis which include, but are not limited to, surge in specific enzymatic activity, shift toward acidic pH, reductive or oxidative states, or a buildup of reactive oxygen species. These homeostatic disturbances can be exploited for the development of targeted therapies that can be activated under certain conditions to trigger drug release. Being aware of these intracellular and extracellular changes allows us to design smart polymer microcapsules and nanocapsules. In addition to these homeostatic disturbances, external physical parameters, such as temperature, ultraviolet light, ultrasound, or magnetic energy, can also be used to trigger drug release from polymer capsules. Thus, in this chapter, we have attempted to summarize the effects of adding biologically responsive moieties to the polymer structure in order to achieve more targeted controlled therapeutic outcomes. They are exogenous and endogenous factors.

4.1 Exogenous factors

Exogenous stimuli result in manipulation of capsule structure from outside the body, such as heat, light, or ultrasound induction.

4.1.1 Temperature sensitive

Temperature sensitive polymeric capsules have been designed to release the drug payloads at the target site due to an induced change in temperature acting as a trigger. The polymers are selected or designed so as to change their physical and chemical properties in response to temperature change. Most tumors and inflammatory conditions tend to exhibit a localized rise in temperature as a result of increased blood flow due to vasodilation or angiogenesis, leukocytic infiltration, increased metabolic activity, and increased cell proliferation and high cell turnover in pathological tissues. Besides this inherent temperature differential between healthy and diseased tissue, external sources can be employed to induce a localized temperature change.

Polymers have been discovered that undergo conformational changes upon temperature variation. These conformational changes could result from a change in hydration states. A majority of temperature sensitive polymers are hydrophilic below a certain temperature and are hydrophobic above a specific temperature. The temperature at which this phase transition occurs is called the lower critical solution temperature (LCST) [12]. The temperature range at which the capsule releases the cargo can be tuned by modulating the balance between hydrophilic and hydrophobic groups in the polymer. It is desirable for a temperature-sensitive capsule to release its cargo at a temperature range of 37–42°C. It is considered to be the optimum working range to attain maximum physiological benefit without any toxic affects due to protein denaturation above 42°C.

Poly(N-isopropylacrylamide) (PNIPAAm) is the most extensively investigated temperature sensitive polymer. It exhibits a LCST of approximately 33°C in aqueous solution, below which PNIPAAm is water soluble and above which it becomes water insoluble. The LCST of PNIPAAm-based polymers can be tuned by copolymerization with hydrophilic or hydrophobic moieties. Incorporation of hydrophilic moieties increases the LCST of the resultant PNIPAAm copolymer and incorporating a hydrophobic moiety would decrease the LCST [13]. However, questions still linger over the biocompatibility of PNIPAAm, which make it an unsavory choice based on the known neurotoxicity of acrylamide monomer and the hydrolysis of PNIPAAm under acidic condition, which yield highly toxic amine molecules [14].

Poly(N-vinylcaprolactam) (PNVCL) is a temperature sensitive polymer that provides a favorable alternative. In contrast to PNIPAAm, PNVCL is biocompatible with extremely low cytotoxicity and does not release toxic small organic amine compounds upon hydrolysis. PNVCL was well tolerated with the Caco-2 and Calu-3 cell lines up to 10 mg mL^{-1}. Further *in vivo* studies are needed to study organ toxicity and elimination to give an in-depth evaluation on its potential. PNVCL-containing materials have been scrutinized for cell immobilization, tissue engineering, anticancer drug delivery, protein separation, etc.

Aliphatic polyesters are attractive thermoresponsive biocompatible and degradable materials for the development of nanocapsules. ε-Caprolactones are biocompatible temperature-sensitive aliphatic polyesters that exhibit a slow biodegradability profile. Cheng et al. [15] synthesized a thermoresponsive poly {γ-2-[2-(2-methoxyethoxy)-ethoxy]ethoxy-ε-caprolactone}-b-poly(γ-octyloxy-ε-caprolactone) (PMEEECL-b-POCTCL) diblock copolymer with a LCST of 38°C. It combined the biocompatibility and biodegradability properties of polycaprolactone with the thermo-responsive properties of oligoethylene glycol-substituted polymers. The thermally sensitive diblock copolymer PMEEECL-b-POCTCL was able to sense the elevated local temperature due to faster metabolism or induced hyperthermia and then release Nile Red and Doxorubicin in a controlled manner (**Figures 1** and **2**). The PMEEECL-b-POCTCL block copolymer also showed relatively low cytotoxicity. Rainbolt et al. [16] synthesized amphiphilic diblock copolymers comprising poly[γ-(2-methoxyethoxy)-Ɛ-caprolactone] and thermosensitive poly{γ-2-[2-(2-methoxyethoxy)ethoxy]ethoxy-Ɛ-caprolactone} (PMECL) as the hydrophobic and hydrophilic blocks, respectively. By controlling the ratio between the monomers MEEECL and MECL, and by varying the number of pendant ethylene oxide units, they were able to achieve highly tunable LCSTs in the range of 31–43°C. PMEEECL-b-PMECL copolymers possessed fully biodegradable backbones. The fundamental lack of chemical functionality in the parent aliphatic polyesters makes it difficult to modify the polymer backbone.

Aliphatic polycarbonates have been employed to produce thermoresponsive biocompatible and degradable materials. Kim et al. [17] synthesized thermoresponsive block copolymers by the ring opening polymerization of cyclic carbonate monomers

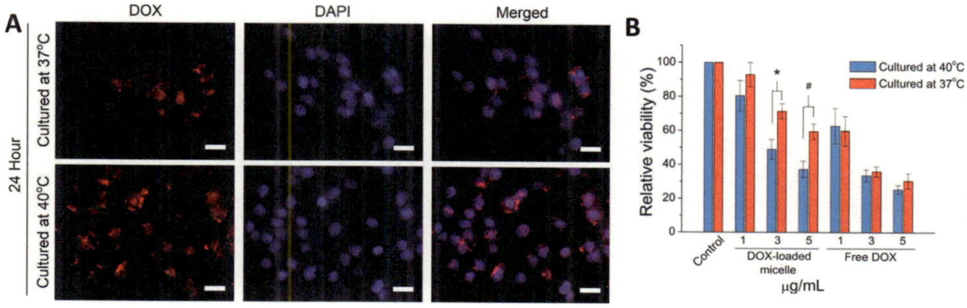

Figure 1.
*(A) CLSM images of MCF-7 cells treated with DOX-loaded PMEEECL-b-POCTCL micelles at either 37 (below LCST) or 40°C (above LCST) for 24 h. For each panel, images from left to right show cells with DOX (red fluorescence), DAPI (blue fluorescence), and overlays of both images. The scale bar indicates 20 μm. (B) The viability of MCF-7 cells incubated with various concentrations of DOX-loaded PMEEECL-b-POCTCL micelles and free DOX at either 37 or 40°C after 24 h of incubation (n = 6). * and # indicate a significant difference between the two participating groups (p < 0.05). Reprinted with permission from Ref. [8]. Copyright 2012 American Chemical Society.*

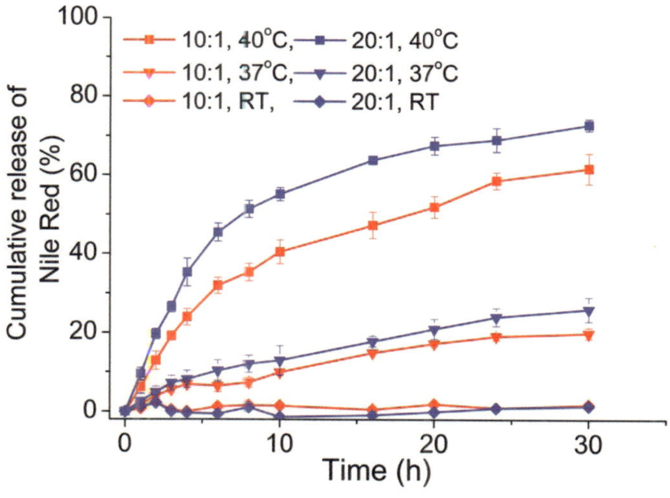

Figure 2.
Cumulative Nile Red release from PMEEECL-b-POCTCL micelle at room temperature, 37 (below LCST) and 40°C (above LCST) in PBS. Reprinted with permission from Ref. [8]. Copyright 2012 American Chemical Society.

functionalized with hydrophilic and hydrophobic groups. The methyltrimethylcarbonate (MTC) family of cyclic carbonates derived from 2,2-bis(methylol)propionic acid (bis-MPA) was exploited as a synthetic building block for functional biodegradable monomers. These polycarbonates derived from cyclic carbonates can be prepared with a range of pendant functional groups that requires only a single functionalization reaction and a simple purification step. Polycarbonates are stable *in vitro*, while they degrade enzymatically *in vivo*. The LCST of these copolymers varied in the range of 36–60°C, depending on the molecular weight of hydrophilic poly(ethylene glycol) (PEG) chains, compositions of copolymers and molar ratios of hydrophilic to hydrophobic monomers. TRC350-10,30,60, (thermoresponsive polycarbonate block copolymer where 10, 30, and 60 refers to degrees of polymerization of MTC-C_2, MTC-PEG and MTC-C_{12}, respectively) which possessed a LCST of 36°C, was identified as a useful model polymer with higher Paclitaxel release at the body temperature (37°C) as compared to a temperature below the LCST. Ajiro et al. [18] developed a thermosensitive biodegradable homopolymer with a poly(trimethylene carbonate) (PTMC) backbone and oligoethylene glycol (OEG). The LCST ranged from 31 to 35°C and was influenced by the molecular weight and polymer concentration.

Some of the other commonly studied thermoresponsive polymers that are focused on LCST-typed polymers are typically based on poly(N,N-diethylacrylamide), poly(methyl vinyl ether) (PMVE), poly(2-ethyl-2-oxazoline) (PEtOx), poly(ethylene oxide)-poly(propylene oxide)-poly(ethylene oxide) (PEO-PPO-PEO), etc.

4.1.2 Ultrasound sensitive

Ultrasound (U/S) is defined as high-frequency pressure waves produced by mechanical oscillations in response to an alternating current applied across piezoelectric materials. Ultrasound devices operate with frequencies from 20 kHz up to several gigahertz. Ultrasonic devices are used to detect objects and invisible flaws and measure distances. Ultrasound is non-invasive and can penetrate deep into tissue of the body. Low-frequency U/S does not damage or heat the tissues, though it is difficult to focus because it produces a larger area of focus. High-frequency U/S can be focused on smaller areas. A high-intensity focused ultrasound (HIFU) beam can target tumors and can harmlessly penetrate the skin and other tissues. Thus, HIFU can be used to treat a variety of tumors.

Ultrasound causes thermal and non-thermal effects on interaction with organic tissue. The hyperthermia is due to the absorption of acoustic energy by fluids and tissues. Thus, U/S can heat the drug carriers, the drugs, and the treated tissues. Hyperthermia induced by the application of U/S is used as an adjuvant in cancer chemotherapy and in physical therapy.

The non-thermal effects of U/S are related to cavitating bubbles. Acoustic cavitation is the interaction of acoustic waves and gas bubbles. Ultrasound reacts with air bubbles and makes them oscillate in response to the negative and positive pressure cycles. The air bubbles oscillate and undergo compression and rarefaction at low acoustic amplitudes, and the frequency of their oscillation resonates with the frequency of the applied U/S. An increase in acoustic pressure results in nonlinear and violent oscillations that ultimately collapse the bubbles (inertial or transient cavitation). The collapse of bubbles and resultant high shear force was demonstrated to enhance drug uptake by cells (*in vivo* and *in vitro*), by temporarily increasing the membrane permeability, and can also dismantle nanocapsules, triggering the release of therapeutic payloads [19].

Husseini et al. [20] extensively studied Pluronic P105 micelles and their release kinetics in relation to U/S exposure. They studied properties like drug release kinetics, pluronic, and drug concentration, the duration of exposure, frequency, intensity, and power density of U/S used. Drug release (doxorubicin and ruboxyl) was most efficient at 20 kHz ultrasound and the data suggests an important role of transient cavitation in drug release. Zhang et al. [21] developed amphiphilic block copolymer PLA-b-PEG encapsulating hydrophobic Nile Red dye. High intensity focused ultrasound (HIFU), as a non-contact and remote control approach, can trigger the release of the encapsulated cargo. The release behavior of Nile Red can be tuned by adjusting the HIFU time, intensity, and location or reactor. An irreversible release due to the degradation of PLA-b-PEG chain resulting from the transient cavitation in the HIFU focal spot was proposed.

2-Tetrahydropyranyl methacrylate (THPMA) is a mechanolabile functional group that can hydrolyze in response to U/S exposure. A novel approach was demonstrated by Xuan et al. [22] to amplify the effect of HIFU on the disassembly of block copolymers (BCP). By introducing a small amount of ultrasound-labile functional group into the thermosensitive block copolymer, the ultrasound-induced reaction of THMPA increased the LCST due to a polarity enhancement. Thus, the BCP becomes soluble in water and results in the disassembly without any changes in the solution temperature. The validity of this new mechanism was shown by synthesizing and investigating a diblock copolymer of PEO_{112}-b-P(MEO_2MA_{189}-co-$THPMA_{21}$). A ^{13}C NMR spectral analysis provided critical evidence to show that the hydrolysis of THPMA groups occurs under HIFU irradiation and that the BCP disassembly originates from an increase in the LCST due to the ultrasound-induced conversion of hydrophobic comonomer units of THPMA onto hydrophilic methacrylic acid (MAA). This general approach of modulating the LCST by ultrasound can be applied by further exploring other ultrasound-labile moieties in the block copolymer design.

The ester bonds and disulfide bonds have been known to have U/S responsiveness. The disulfide bond has a relatively low dissociation energy ($E_{S–S}$ ~268 kJ mol^{-1}) and longer bond length ($l_{S–S}$ B 2.03 Å) compared with those of the C–C bond ($E_{C–C}$ ~347 kJ mol^{-1}; $l_{C–C}$ B 1.54 Å) [23]. Li et al. [24] synthesized biodegradable block copolymer (PEG-S-S-PLA) containing the central labile disulfide linkage between polyethylene glycol (PEG) and poly-L-lactic acid (PLA) segments. When the BCP are subjected to HIFU, the labile disulfide bonds are cleaved, and the amphiphilic structures are broken, which further leads to the disruption of BCP and the release of the encapsulated cargo. HIFU disrupts PEG-S-S-PLA structure because of the HIFU-induced site specific degradation of PEG-S-S-PLA.

Tong et al. [25] synthesized pluronic type copolymer PEG-COO-SS-PPG containing disulfide and mechanolabile ester bonds at the junction points of PEG and PPG blocks. They demonstrated that HIFU caused polymer degradation of the BCP and was substantiated by decrease in molecular weight. The molecular weight of PEG-COO-SS-PPG decreases with increase in HIFU exposure time. They also concluded that the cavitation was the primary reason for the cleavage of ester bonds and polymer degradation, and not the U/S-induced thermal effect. The disulfide bonds in the copolymer PEG-COO-SS-PPG possessed redox responsive property. Thus, the copolymer featured ultrasound and redox dual responsiveness exhibited relatively slow redox-induced release behavior and fast HIFU-induced release behavior. Ultrasound triggered mechanochemical cleavage occurred preferentially at the central ester bond rather than at the disulfide bond.

Wang et al. [26] developed a diblock copolymer composed of poly(ethylene oxide) and poly(2-tetrahydropyranyl methacrylate) (PEO-b-PTHPMA) that could be hydrolyzed by high-frequency ultrasound (1.1 MHz). HIFU beam could induce the hydrolysis reaction of THPMA at room temperature resulting in the cleavage of THP groups leading to the formation of carboxylic acid dimers and hydroxyl groups. As shown in **Figure 3**, it was found that upon exposure to a high-intensity focused ultrasound (HIFU) beam at room temperature, the pH value decreased over irradiation time. It was also found that with increase in ultrasound power output, the pH decrease was more significant, thus demonstrating ultrasound triggered polymer degradation.

Ultrasound triggered drug release has many advantages. It provides the ability to control the time of drug release from the carrier. This is especially desirable when it is required to release the majority of the drug instantly and simultaneously to achieve a rapid high and lethal concentration. It would also be prudent to keep in mind that whenever ultrasound is utilized, its parameters, such as the power density, acoustic frequency, continuous or pulsed U/S, and pulse duration, have to be optimized to reach the desired therapeutic effect while minimizing the damage to the adjacent health tissues [19].

4.1.3 Light sensitive

Capsules capable of undergoing physical or chemical changes in response to light irradiation offer spatiotemporal control over the release of encapsulated therapeutic payloads. Chemical and physical processes that can be initiated by light irradiation at a specific wavelength can be either reversible or irreversible. These processes can involve formation or cleavage of bonds, interconversion of isomers, electrostatic charge switching, and rearrangement of chemical reactions. Light-triggered polymeric capsules can be developed by the incorporation of functional groups that interact with light which are triggered to disintegrate and release the encapsulated payload via light irradiation.

The suitability of these capsules for any biomedical application is dependent on the radiation wavelength required. The radiation required should be benign to live tissues, exhibit minimal absorption, and interaction with the biological components and offer substantial tissue penetration for *in vivo* applications. X-Rays and γ-rays have short wavelengths and possess high energy that would damage normal tissues, making them unsuitable for this purpose. UV radiation with its broader wavelength and less energy is more suited for developing light-sensitive capsules. Azobenzene and spiropyran have been known to undergo UV light-triggered photoisomerization. Incorporation of these functional groups into the polymer chain (in the backbone or as pendant groups) has helped to develop UV light-triggered capsules.

Under UV light radiation, Azobenzene undergoes photoisomerization from apolar *trans* to polar *cis* isomeric form. This isomeric transition can be reversed by

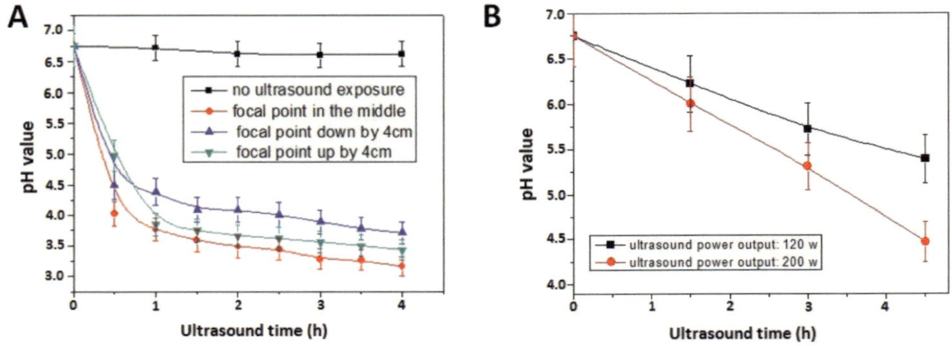

Figure 3.
(A) Change in pH as a function of ultrasound irradiation time. By repositioning the tube reactor, the location of the focal point of the ultrasound beams can be varied (ultrasound power, 200 W). (B) Change in pH as a function of ultrasound irradiation time for PEO-b-PTHPMA, using two different ultrasound power outputs. Reprinted with permission from Ref. [27]. Copyright 2009 American Chemical Society.

storage in the dark or by irradiation with visible light. The change in hydrophilicity and the transition from *trans* to *cis* conformation can cause disintegration of the nanocapsules derived from polymers bearing azobenzene groups, leading to payload release.

Xiao et al. [28] developed photo-switchable microcapsules based on host-guest interactions between α-cyclodextrin (α-CD) and azobenzene. Confocal laser scanning microscopy was used to observe the photo-dissociation of capsules irradiated by UV light, as seen in **Figure 3**. These microcapsules exhibited controlled release behavior. Blasco et al. [29] developed linear-dendritic block copolymers which formed polymeric vesicles that are triggered by low intensity UV light, therefore limiting the possible toxic effects to organic tissue when exposed to UV radiation. The cyano group at the para-position of the azobenzene moiety was substituted by an alkyloxy group. This modification was instituted to increase the polarity between *trans* and *cis* isomers. They demonstrated that structural modification with 4-isobutyloxyazobenzene incorporation facilitated the disruption of azobenzene aggregates of the membrane on exposing the vesicles to low intensity UV light when compared to its 4-cyanoazobenzene and azobenzene counterparts. Blasco et al. [30] continued this work to study the photo-responsiveness of these polymer vesicles with different percent of 4-isobutyloxyazobenzene (IBO) and hydrocarbon chains (C18) randomly distributed at the periphery of the dendron. Results indicated that by diluting the azobenzene content at the periphery of the dendron, the *trans*-to-*cis* photoisomerization rate can be substantially accelerated and the light-induced release activity can be tuned. Vesicles with a 50/50 IBO/C18 ratio suffered the most significant changes upon UV irradiation. Yi and co-workers [31] developed a new class of UV responsive polyelectrolyte microcapsules. Upon exposure to UV light, the azobenzene moieties in the multilayers self-organized in the form of J aggregates due to the influence of polycation [poly(diallyldimethyl ammonium) chloride, (PDADMAC)]. The SEM images presented in **Figure 4** revealed that the re-orientation of azobenzene within

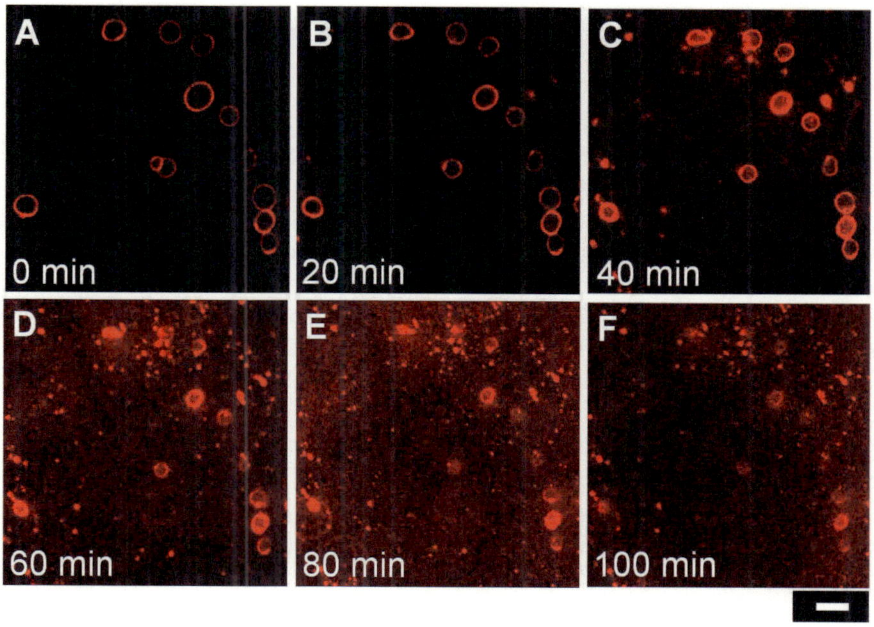

Figure 4.
Snapshots of the photodissociation of microcapsules (A)–(F). The time interval between the snapshots is 20 min. Scale bar 15 μm. Reprinted with permission from Ref. [19]. Copyright 2011 American Chemical Society.

shell formations led to great damage of capsule integrity, illustrating the progress of capsules swelling and their disruption further. After 2 h of UV irradiation, the capsule debris was split into needle-like formations. This UV-induced microcapsule disruption process was proven to be irreversible.

Among the class of synthetic photoresponsive molecules, spiropyran (SP) has been known to possess unique tunability, stability, and fast response time. Spiropyran can undergo a reversible transformation from colorless to pink-colored merocyanine (MC) upon UV irradiation with a marked increase in the polarity associated with the structural conversion from neutral to charge-separated zwitterions [32]. This solubility switching of spiropyran from hydrophobic to hydrophilic can induce the destabilization of polymeric nanocapsules. Moreover, this destabilization induces controlled release of model hydrophobic therapeutics by UV irradiation. The transition can be reversed by irradiation with visible light. Achilleos and coworkers [33] developed water-dispersible nanocapsules based on the formation of H-type π–π interactions between the merocyanine (MC) isomers within the sterically crowded environment of the polymer brushes upon UV irradiation, which enables the SP-to-MC isomerization of the photosensitive species. Disruption of the nanocapsules can be achieved remotely by applying a harmless trigger such as visible-light irradiation.

Besides azobenzene and spiropyran, drug carriers based on cinnamic acid, cinnamic ester, and coumarin are also capable of responding to UV irradiation. Additionally, 2-diazo-1,2-naphthoquinone (DNQ), o-nitrobenzyl ester, coumarinyl ester, and pyrenylmethyl ester groups, which undergo cleavage, have also been explored for the development of UV-radiation-responsive drug carriers [34].

4.2 Endogenous factors

Endogenous stimuli are generated as a result of inherent chemical biological pathology in diseased states, such as pH, reactive oxygen species, and elevated enzyme levels.

4.2.1 pH sensitive

pH variations between different tissue compartments have been exploited as a shining platform for the development of pH-sensitive drug-delivery systems. Normal extracellular tissue and systemic blood have a physiological pH of 7.4. However, tumor tissues having an abnormally high cell proliferation rate and the lack of nutrients leads to a high rate of glycolysis and accumulation of lactic acid, which lower the environmental pH by 0.5–1.0 unit as compared to healthy tissues. The intracellular organelles, like endosomes and lysosomes, have a more acidic pH of 4.0–6.5 [35]. This pH gradient can be exploited as an endogenous stimulus for the development of nanocapsules to selectively deliver and activate drug molecules while reducing their systemic side-effects. pH-responsive nanocapsules can be programmed by functionalizing the polymer backbone with hydrazones, amines, acetals, ketals, boronic acid, oximes, etc.

Literature indicated that hydrazone is the most extensively explored pH-responsive chemical bond due to its sharp responsiveness to pH changes. Ganivada et al. [36] synthesized a newly designed biodegradable copolymer (PVLPEG-PVLDOXI-PCL-PHOS-Fe_3O_4) that can be used as a nanocapsule. The magnetic nature of Fe_3O_4 is expected to behave as a smart nanocarrier for both magnetic resonance imaging (MRI) as well as efficient sustained delivery vehicle for doxorubicin. The acylhydrazine linker helps to release the drug under mild acidic conditions. Wang et al. [37]

developed a biodegradable endosomal pH-sensitive polymeric prodrug based on poly(5-methyl-5-allyloxycarbonyl-1,3-dioxan-2-one)-*graft*-12-acryloyloxy dodecyl phosphorylcholine (PMAC-*graft*-ADPC). Doxorubicin was conjugated to the polymer through hydrazone bonds. *In vitro* drug release studies showed that the release of doxorubicin was faster at endosomal pH (pH = 5.0) than at normal physiological environment (pH = 7.4), as shown in **Figure 5**.

Acetal is another acid-labile functional group that can be wielded to design smart polymeric nanocapsules. Tonhauser et al. [38] developed long chain branched and hyperbranched polyether polyols by copolymerization of an acetal-containing inimer, namely, 1-(glycidyloxy)ethyl ethylene glycol ether (GEGE) with ethylene oxide and glycidol. Owing to the presence of acetal group, a strong pH-dependence of the degradation kinetics was observed. At pH 4.5, $t_{1/2}$ (half-life) is approximately 76 h, while at pH 4, $t_{1/2}$ is almost one-third of this value with 26 h. Also, since no degradation was observed at pH 7 or higher, storage stability was guaranteed. Hu et al. [39] designed a series of well-defined three-armed star-block copolymers containing poly(ethylene glycol) monomethyl ether (mPEG) and poly(ε-caprolactone) blocks linked with acid-cleavable acetal groups to demonstrate the pH-dependent release of doxorubicin. Tomlinson et al. [40] proved that the pH responsiveness of acetal-labile polymer was due to polymer degradation. It was substantiated by a pH-dependent decrease in molecular weight at pH 7.4, 6.5, and 5.5.

β-Carboxylic amide is another acid-sensitive functional group that can be included as a pendant chain to incorporate its acid-sensitivity to the parent polymer chain. β-Carboxylic amides functionalized polymer chains are negatively charged and stable at neutral pH, but when the pH is decreased to 6, they quickly become positively charged due to the hydrolysis of β-carboxylic amides in acidic condition. After hydrolysis, the β-carboxylic amides form cationic primary amines. This charge reversal or charge conversion can have two main effects. In 1997, Behr [41] proposed the "proton-sponge" hypothesis describing that unprotonated amines can absorb protons as they are pumped into the lysosome, resulting in more protons being pumped in leading to an increased influx of chlorine ions and water. This increases endosomal pH. A combination of osmotic swelling and swelling of the amine-functionalized polymer triggers endosomal escape, with subsequent release of its contents into the cytoplasm because of repulsion between protonated amine groups, which causes the rupture of the lysosomal membrane. Secondly, β-carboxylic amides modified

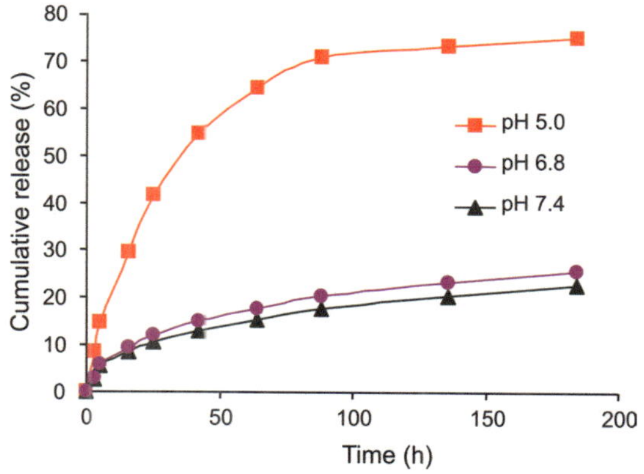

Figure 5.
Time-dependent cumulative release of doxorubicin from PPC-Hyd-DOX-DA NPs at different pH values. Reprinted with permission from Ref. [33]. Copyright 2011 American Chemical Society.

polymer can exploit its negative charge interaction with cationic drugs like doxorubicin to achieve high drug-loading capacity and efficiency. The pH induces charge reversal in acidic condition, and generates a repulsive electrostatic force between positively charged amines and cationic drugs, leading to accelerated drug release. Quadir et al. [42] showed that PEG–polypeptide copolymers with amine pendant chains can be utilized as a pH-responsive drug-delivery vehicle. These nanosized vesicles were proved to be stable platforms for drug delivery that can successfully localize in solid tumors and release doxorubicin, mediated by the enhanced permeation, and retention effect by lowering the pH of the tumor microenvironment.

The pH difference between healthy and disease tissue, and between extracellular and intracellular compartments, has been exploited to fabricate the pH-responsive polymer nanocapsules for controlled drug release. This field is the target of extensive investigations being done worldwide.

4.2.2 Reactive oxygen species sensitive

Reactive oxygen species (ROS) are partially reduced chemically reactive metabolites of oxygen. These include superoxide anion, hydroxyl radical, hydrogen peroxide, peroxynitrite, and hypochlorous acid. ROS have important roles in cell signaling and homeostasis. However, high levels in certain pathological conditions can damage the cells by causing oxidations of lipids, proteins, and DNA. These have also been known to cause oxidative deactivation of specific enzymes by oxidation of co-factors. This is known as oxidative stress. Elevated levels of ROS have been implicated in a number of pathological conditions, including aging, male infertility, inflammation, infections, cancer, atherosclerosis, reperfusion injury, and diabetes [43]. This has sparked a lot of interest into finding ways to create ROS-eliminating strategies. This has been achieved by two mechanisms.

The first mechanism involves a phase transition of the polymer backbone from hydrophobic to hydrophilic, thus making the capsule wall material soluble in solution and release the therapeutic payload. Sulfides and selenides are the two functional groups, when subjected to ROS, increase their solubility in water. Ma et al. [27] synthesized a diselenide-containing block copolymer and both the oxidizing and reducing groups were shown to have the capability to destroy Se-Se bonds. Poole et al. [44] developed ROS-responsive poly(propylene sulfide) (PPS) microspheres, which were successfully utilized for encapsulation and sustained delivery of curcumin in the diabetic mouse hind limb ischemia model of peripheral arterial disease. The PPS microspheres were biocompatible, improved cell survival under exogenous oxidative stress and reduced ROS levels both *in vitro* and *in vivo*.

The second mechanism involves the physical degradation of polymer backbone in response to ROS. Functional groups like boronic-esters, thioketals, and prolines have been proven to be ROS sensitive. Zhang et al. [45] synthesized 4-phenylboronic acid pinacol ester (PBAP) chemically conjugated onto hydroxyl groups of β-CD to synthesize oxidatively responsive β-cyclodextrin. ROS-sensitivity and superior biocompatibility of this boronic-ester functionalized cyclodextrin were validated via extensive *in vitro* and *in vivo* studies. Lee et al. [46] developed a ROS-degradable scaffold by crosslinking PCL with a ROS-degradable oligoproline peptide, KP_7K, which was further investigated in an *in vivo* model for long-term tissue engineering applications. The KP_7K-crosslinked polymer scaffolds underwent significant surface degradation in response to ROS.

In 2010, Wilson et al. [47] developed thioketal nanoparticles formulated via a polymer, poly-(1,4-phenyleneacetone dimethylene thioketal), that degraded selectively in response to ROS. Thioketal bonds were found to be stable to acid-, base-, and protease-catalyzed polymer degradation. In a murine model of ulcerative colitis, orally administered thioketal nanoparticles, loaded with siRNA against the proinflammatory cytokine TNF-α, remained stable in the harsh environment of the gastrointestinal tract, protecting TNF–siRNA and preventing its release to non-inflamed mucosal tissues. However, at sites of intestinal inflammation, where infiltrating phagocytes produce unusually high levels of ROS, the thioketal bond crosslinked polymer degrades thus releasing TNF-siRNA to the site of inflammation. This led to controlled gene silencing to curtail TNF-mRNA levels in the colon and protect the mice from ulcerative colitis.

4.2.3 Enzyme sensitive

Enzymes are the key biological catalysts that play a vital role in all chemical, metabolical, and biological processes serving as the prime protagonists in the chemistry of living organisms at a molecular level. As their presence and activities are essential in maintaining the physiological homeostasis, enzyme dysregulation is associated with many diseases and pathological disorders, such as cancer, cardiovascular disease, inflammation, osteoarthritis, Alzheimer's disease, inborn errors of metabolism, etc. Therefore, altered expression level of specific enzymes can be exploited as a pristine biological trigger to achieve enzyme-mediated biomaterial responses and controlled release of biomolecules at the desired sites. Compared to typical catalyzed or non-catalyzed chemical reactions involved in polymer synthesis, enzyme-catalyzed reactions exhibit superior advantages, such as high selectivity and substrate specificity. These reactions can be conducted under quite mild condition (37°C, aqueous media, typically neutral or slightly acidic and alkaline pH). Another advantage is that enzyme-catalyzed reactions can be conducted *in vitro*.

Azobenzene is an artificial structural moiety that can be used to fabricate azoreductase-sensitive nanocapsules. The enzyme azoreductase is produced by microbial flora specifically present in the colon of the human intestine which makes it an elegant stimulus to create colon-specific drug-delivery systems. Rao et al. [48] developed an amphiphilic diblock copolymer with an azobenzene linkage introduced as an artificial enzyme active site at the junction of the diblock copolymer. The authors documented the enzymatic dissociation of the copolymer connection which released the two polymer segments following introduction of the enzyme azoreductase, in the presence of coenzyme Dihydronicotinamide-adenine dinucleotide phosphate (NADPH), leading to the cleavage of the azobenzene-based block copolymer linkage. The authors thus established the potential applications of azobenzene linkages in the realm of colon-specific drug-delivery systems.

Gelatinase is a mixture of two types of matrix metalloproteinase (MMP) enzymes that is highly expressed in tumor tissues. Dong et al. [49] developed a pH and enzyme-responsive complex composed of doxorubicin, CpG DNA fragments, cationic gelatin, and pH-sensitive alginate. Since gelatinase is relatively highly expressed in liver, a PEGylated pH-responsive alginate was introduced to reduce the liver accumulation of doxorubicin and thus attenuate its hepatotoxicity. The gelatinase and DNase present in high concentration in the cytoplasm of tumor cells helped to dissociate the cationic gelatin and CpG DNA fragments and enhanced the drug release.

5. Future Prospects and Challenges

Wall materials are already playing an important role in the development of micro-/nanoformulations, which have been applied as drug-delivery systems. These drug-delivery systems have greater potential for many applications, including anti-tumor therapy, gene therapy, AIDS-therapy, radiotherapy, delivery of proteins, antibiotics, virostatics, vaccines, and as vesicles to cross the blood-brain barrier. Nanocapsules particularly provide massive advantages regarding drug targeting, delivery, and release. With their additional potential to combine diagnosis and therapy, they emerge as one of the major tools in nanomedicines. The major challenges are to improve their stability in a biological environment, to mediate the biodistribution of active compounds, improve drug loading, and establish greater interaction with the biological barriers. The cytotoxicity and the degradation products of the formulations developed from the wall materials remain a major problem. Thus, improvements in biocompatibility are the main concern of future research. These challenges can be suitably addressed by developing newer polymeric wall materials, which can be of better biocompatibility with the biological environment. Simultaneously, molecular modeling study would greatly contribute in understanding the interaction between the wall materials and drug, and this in turn with the biological environment. Finally, the success of this is largely dependent on the efforts of scientists, physicians, and engineers.

Acknowledgements

The authors thank the University Grants Commission (UGC), New Delhi and Department of Science and Technology (DST), New Delhi for providing the financial support respectively under UPE-FAR-I program [Grant No. 14-3/2.12(NS/PE)] and DST-PURSE-Phase-II program [Grant No. SR/PURSE PHASE-2/13(G)].

Author details

Shaluah Vijeth, Geetha B. Heggannavar, Mahadevappa Y. Kariduraganavar*
Department of Chemistry, Karnatak University, Dharwad, India

*Address all correspondence to: mahadevappayk@gmail.com

References

[1] Bansode SS, Banarjee SK, Gaikwad DD, Jadhav SL, Thorat RM. Microencapsulation: A review. International Journal of Pharmaceutical Sciences Review and Research. 2010;**1**:38-43

[2] Lohcharoenkal W, Wang L, Chen YC, Rojanasakul Y. Protein nanoparticles as drug delivery carriers for cancer therapy. BioMed Research International. 2014;**2014**:1-12. Article ID: 180549

[3] Elgadir MA, Uddin MS, Ferdous S, Adam A, Chowdhury AJK, Sarker MZI. Impact of chitosan composites and chitosan nanoparticle composites on various drug delivery systems: A review. Journal of Food and Drug Analysis. 2014;**23**:619-629

[4] Schnürch AB, Dünnhaupt S. Chitosan-based drug delivery systems. European Journal of Pharmaceutics and Biopharmaceutics. 2012;**81**:463-469

[5] Hu L, Suna Y, Wu Y. Advances in chitosan-based drug delivery vehicles. Nanoscale. 2013;**5**:3103-3111

[6] Tripodo G, Trapani A, Torre ML, Giammona G, Trapani G, Mandracchia D. Hyaluronic acid and its derivatives in drug delivery and imaging: Recent advances and challenges. European Journal of Pharmaceutics and Biopharmaceutics. 2015;**97**:400-416

[7] Dosio F, Arpicco S, Stella B, Fattal E. Hyaluronic acid for anticancer drug and nucleic acid delivery. Advanced Drug Delivery Reviews. 2016;**97**:204-236

[8] Zhang H, Huang S, Yang X, Zhai G. Current research on hyaluronic acid-drug bioconjugates. European Journal of Medicinal Chemistry. 2014;**86**:310-317

[9] Pohlmann AR, Fonseca FN, Paese K, Detoni CB, Coradini K, Beck RC, et al. Poly(ϵ-caprolactone) microcapsules and nanocapsules in drug delivery. Expert Opinion on Drug Delivery. 2013;**10**:623-638

[10] Mirakabad FST, Koshki KN, Akbarzadeh A, Yamchi MR, Milani M, Zarghami N, et al. PLGA-based nanoparticles as cancer drug delivery systems. Asian Pacific Journal of Cancer Prevention. 2014;**15**:517-535

[11] Suka JS, Xua Q, Kima N, Hanesa J, Ensigna LM. PEGylation as a strategy for improving nanoparticle-based drug and gene delivery. Advanced Drug Delivery Reviews. 2016;**99**:28-51

[12] Roy D, Brooks WLA, Sumerlin BS. New directions in thermoresponsive polymers. Chemical Society Reviews. 2013;**42**:7214-7243

[13] Wei H, Cheng SX, Zhang XZ, Zhuo RX. Thermo-sensitive polymeric micelles based on poly(N-isopropylacrylamide) as drug carriers. Progress in Polymer Science. 2009;**34**:893-910

[14] Liu J, Debuigne A, Detrembleur C, Jérôme C. Poly(N-vinylcaprolactam): A thermoresponsive macromolecule with promising future in biomedical field. Advanced Healthcare Materials. 2014;**3**:1941-1968

[15] Cheng Y, Hao J, Lee LA, Biewer MC, Wang W, Stefan MC. Thermally controlled release of anticancer drug from self-assembled γ-substituted amphiphilic poly(ε-caprolactone) micellar nanoparticles. Biomacromolecules. 2012;**13**:2163-2173

[16] Rainbolt EA, Washington KE, Biewer MC, Stefan MC. Towards smart polymeric drug carriers: Self-assembling γ-substituted polycaprolactones with highly tunable thermoresponsive behavior. Journal of Materials Chemistry B. 2013;**1**:6532-6537

[17] Kim SH, Tan JPK, Fukushima K, Nederberg F, Yang YY, Waymouth RM, et al. Thermoresponsive nanostructured polycarbonate block copolymers as biodegradable therapeutic delivery carriers. Biomaterials. 2011;**32**:5505-5514

[18] Ajiro H, Takahashi Y, Akashi M. Thermosensitive biodegradable homopolymer of trimethylene carbonate derivative at body temperature. Macromolecules. 2012;**45**:2668-2674

[19] Ahmed SE, Martins AM, Husseini GA. The use of ultrasound to release chemotherapeutic drugs from micelles and liposomes. Journal of Drug Targeting. 2015;**23**:16-42

[20] Husseini GA, Myrup GD, Pitt WG, Christensena DA, Rapoporta NY. Factors affecting acoustically triggered release of drugs from polymeric micelles. Journal of Controlled Release. 2000;**69**:43-52

[21] Zhang H, Xia H, Wang J, Li Y. High intensity focused ultrasound-responsive release behavior of PLA-b-PEG copolymer micelles. Journal of Controlled Release. 2009;**139**:31-39

[22] Xuan J, Boissière O, Zhao Y, Yan B, Tremblay L, Lacelle S, et al. Ultrasound-responsive block copolymer micelles based on a new amplification mechanism. Langmuir. 2012;**28**:16463-16468

[23] Skinner HA. A revision of some bond-energy values and the variation of bond-energy with bond-length. Transactions of the Faraday Society. 1945;**41**:645-662

[24] Li Y, Tong R, Xia H, Zhang H, Xuan J. High intensity focused ultrasound and redox dual responsive polymer micelles. Chemical Communications. 2010;**46**:7739-7741

[25] Tong R, Lu X, Xia H. A facile mechanophore functionalization of an amphiphilic block copolymer towards remote ultrasound and redox dual stimulus responsiveness. Chemical Communications. 2014;**50**:3575-3578

[26] Wang J, Pelletier M, Zhang H, Xia H, Zhao Y. High-frequency ultrasound-responsive block copolymer micelle. Langmuir. 2009;**25**:13201-13205

[27] Ma N, Li Y, Xu H, Wang Z, Zhang X. Dual redox responsive assemblies formed from diselenide block copolymers. Journal of the American Chemical Society. 2010;**132**:442-443

[28] Xiao W, Chen WH, Zhang J, Li C, Zhuo RX, Zhang XZ. Design of a photoswitchable hollow microcapsular drug delivery system by using a supramolecular drug-loading approach. The Journal of Physical Chemistry, B. 2011;**115**:13796-13802

[29] Blasco E, Barrio JD, Somolinos CS, Pinol M, Oriol L. Light induced molecular release from vesicles based on amphiphilic linear-dendritic block copolymers. Polymer Chemistry. 2013;**4**:2246-2254

[30] Eva Blasco E, Serrano JL, Pinol M, Oriol L. Light responsive vesicles based on linear-dendritic block copolymers using azobenzene–aliphatic codendrons. Macromolecules. 2013;**46**:5951-5960

[31] Yi Q, Sukhorukov GB. UV-induced disruption of microcapsules with azobenzene groups. Soft Matter. 2014;**10**:1384-1391

[32] Lee HI, Wu W, Oh JK, Mueller L, Sherwood G, Peteanu L, et al. Light-induced reversible formation of polymeric micelles. Angewandte Chemie (International Ed. in English). 2007;**46**:2453-2457

[33] Achilleos DS, Hatton A, Vamvakaki M. Light-regulated supramolecular engineering of polymeric nanocapsules. Journal of the American Chemical Society. 2012;**134**:5726-5729

[34] Kamaly N, Yameen B, Wu J, Farokhzad OC. Degradable controlled-release polymers and polymeric nanoparticles: Mechanisms of controlling drug release. Chemical Reviews. 2016;**116**:2602-2663

[35] Pang X, Jiang Y, Xiao Q, Leung AW, Hua H, Xu C. pH-responsive polymer-drug conjugates: Design and progress. Journal of Controlled Release. 2016;**222**:116-129

[36] Ganivada MN, Rao VN, Dinda H, Kumar P, Sarma JD, Shunmugam R. Biodegradable magnetic nanocarrier for stimuli responsive drug release. Macromolecules. 2014;**47**:2703-2711

[37] Wang H, Wang Y, Chen Y, Jin Q, Ji J. A biomimic pH-sensitive polymeric prodrug based on polycarbonate for intracellular drug delivery. Polymer Chemistry. 2014;**5**:854-861

[38] Tonhauser C, Schüll C, Dingels C, Frey H. Branched acid-degradable, biocompatible polyether copolymers via anionic ring-opening polymerization using an epoxide inimer. ACS Macro Letters. 2012;**9**:1094-1097

[39] Hu J, He J, Zhang M, Ni P. Precise modular synthesis and a structure–property study of acid-cleavable star-block copolymers for pH-triggered drug delivery. Polymer Chemistry. 2015;**6**:1553-1566

[40] Tomlinson R, Klee M, Garrett S, Heller J, Duncan R, Brocchini S. Pendent chain functionalized polyacetals that display pH-dependent degradation: A platform for the development of novel polymer therapeutics. Macromolecules. 2002;**35**:473-480

[41] Behr JP. The proton sponge: A trick to enter cells the viruses did not exploit. Chimia International Journal for Chemistry. 1997;**51**:34-36

[42] Quadir MA, Morton SW, Deng ZJ, Shopsowitz KE, Murphy RP, Epps TH, et al. PEG–polypeptide block copolymers as pH-responsive endosome-solubilizing drug nanocarriers. Molecular Pharmaceutics. 2014;**11**:2420-2430

[43] Lee SH, Gupta MK, Bang JB, Bae H, Sung HJ. Current progress in reactive oxygen species (ROS)-responsive materials for biomedical applications. Advanced Healthcare Materials. 2013;**2**:908-915

[44] Poole KM, Nelson CE, Joshi RV, Martin JR, Gupta MK, Haws SC, et al. ROS-responsive microspheres for on demand antioxidant therapy in a model of diabetic peripheral arterial disease. Biomaterials. 2015;**41**:166-175

[45] Zhang D, Wei Y, Chen K, Zhang X, Xu X, Shi Q, et al. Biocompatible reactive oxygen species (ROS)-responsive nanoparticles as superior drug delivery vehicles. Advanced Healthcare Materials. 2015;**4**:69-76

[46] Lee SH, Boire TC, Lee JB, Gupta MK, Zachman AL, Rath R, et al. ROS-cleavable proline oligomer crosslinking of polycaprolactone for pro-angiogenic host response. Journal of Materials Chemistry B. 2014;**2**:7109-7113

[47] Wilson S, Dalmasso G, Wang L, Sitaraman SV, Merlin D, Murthy N. Orally delivered thioketal nanoparticles loaded with TNF–siRNA target inflammation and inhibit gene expression in the intestines. Nature Materials. 2010;**9**:923-928

[48] Rao J, Khan A. Enzyme sensitive synthetic polymer micelles based on the azobenzene motif. Journal of the American Chemical Society. 2013;**135**:14056-14059

[49] Dong L, Xia S, Wua K, Huang Z, Chen H, Chen J, et al. A pH/Enzyme-responsive tumor-specific delivery system for doxorubicin. Biomaterials. 2010;**31**:6309-6316

Chapter 2

Microencapsulated Vegetable Oil Powder

Ekasit Onsaard and Wiriya Onsaard

Abstract

Vegetable oil has been increasingly popular among consumption oils as it provides several health benefits such as antioxidant, anti-inflammatory, antivasoconstrictive, antiarrhythmic, antithrombotic, antimicrobial, antihypertension, antiaging, etc. Several applications of vegetable oils in foods, cosmetics, and pharmaceutical industries have been widely researched as it is made from natural products with a safe and reliable process. However, oxidative deterioration and stabilization of vegetable oil provide short shelf life storage with poor consumer acceptance. Thus, this chapter is aimed to give an overview of stabilization of vegetable oils using microencapsulation techniques mostly focusing on emulsion preparation using multilayer emulsion followed by a spray drying technique to obtain vegetable oil powder. Using different wall materials was discussed along with the application for several vegetable oils. Moreover, the characterization of encapsulated vegetable oil powder was summarized for the final product quality and encapsulated process efficiency.

Keywords: microencapsulation, vegetable oil, multilayer emulsion, spray drying, vegetable oil powder

1. Introduction

Vegetable oils are a group of fats that are extracted from seeds or the other part of a seed plant. In 2018/2019, the world production of major oilseeds is forecasted to 604.67 million metric tons (MMT). Three main oilseed productions are soybean (369.32 MMT), rapeseed (71.70 MMT), and sunflower seed (49.83 MMT) [1]. The main oilseed product exporting countries are USA, Brazil, Canada, China, the European Union, and Argentina. USA soybean has been exported to China in the total of 1.3 million tons compared to 10.7 million tons of fruits or vegetables in the last year [1]. It is recognized that vegetable oils and their products have an important role to make the economy of many countries. These oils are an important source of energy and a carrier of fat-soluble vitamins [2]. Common vegetable oils include soybean, rapeseed, sunflower seed, corn, sesame, coconut, rice, etc. Currently, vegetable oil consumption has been increasing compared to animal fat consumption. Three exporters of oilseeds including copra, cottonseed, palm kernel, peanut, rapeseed, soybeans, and sunflower seeds are Brazil (75.28 MMT), the USA (57.20 MMT), and Canada (17.13 MMT) [1].

Vegetable oils are an important renewable resource from nature containing ester mixtures derived from glycerol with chains of fatty acid about 14–20 carbon atoms with different degree of unsaturation [3]. There are composed of mixtures of triacylglycerols (TAG) (>90–95%) with minor diacylglycerols, tocopherols/tocotrienols, and phytosterol ester (<5–10%). Several health benefits of vegetable oils are gastronomic, nutritional, organoleptic, antioxidant, anti-inflammatory, anti-vasoconstrictive, antiarrhythmic, antithrombotic, antimicrobial, antihypertension, antiaging, etc. [4]. The vegetable oils and their components have been growing of interest in food, cosmetics, and pharmaceutical industries as of their natural and safety produce, and the acceptance by the consumer has been found increasingly. Although vegetable oils have gained popularity and interest, they are sensitive to oxidative deterioration and generate several degradation products such as aldehyde, ketones, epoxides, hydroxyl compounds, etc. These changes occurring in vegetable oil affect shelf life, sensory properties, and overall acceptability of products. Microencapsulation technique has been applied as it has the potential to delay lipid oxidation rate of vegetable oils. Several studied have shown that vegetable oil can play an important role in protection against oxidation using microencapsulation technique [5–10]. Microencapsulation (ME) is the technique in which small particle or liquid droplets are coated or are embedded in a homogenous or heterogonous matrix to form small capsules in both dry form and wet form products [11]. However, there are several methods of encapsulated vegetable oil powder include emulsification, spray drying, freeze drying, fluidized bed coating, extrusion, cocrystallization, molecular inclusion, coaxial electrospray system, and coacervation [4, 12]. Therefore, the objective of this chapter is conveying an overview of the microencapsulated vegetable oil powder method and technique. This chapter summarizes the preparation of vegetable oil-in-water emulsion stabilized by proteins and other wall materials, providing information on microencapsulated powder using spray drying, and characterization of microencapsulated powder and application is finally discussed.

2. Microencapsulation

Microencapsulation (ME) is a technique in which solid, or liquid, gaseous active is coated by a coating material to give small capsules aiming to obtain some physical or chemical properties, which can be applied in a food system. The terminology used to describe "microparticle" refers to a particle with diameter from 1 to 1000 μm, irrespective of the interior or exteriors structure. Generally, nanocapsule refers to a particle range from 10 to 1000 nm [13]. Microspheres refer to spherical microparticles, and subcategory of microcapsules applies to microparticles, which have a core surrounded by materials. "Microcapsule" is defined as a spherical particle size that ranges 50–2 mm including a core material where microspheres are spherically empty particles. However, both microcapsules and microspheres are often used synonymously [13]. In addition, some related terms are used alternatively called "microbeads" and "beads." Moreover, some particles greater than 1000 μm can be termed microgranules or macrocapsules.

Encapsulated material located inside small capsules is known as core materials or internal phase or active ingredient, whereas the outer or protective materials are called as wall material, carrier, shell, or encapsulation matrix (**Figure 1**). The wall protects the core materials from environment such as light, oxygen, moisture, etc. Wall materials can be commonly used as both synthetic polymers and biomaterials (carbohydrates and proteins or combination materials). Therefore, the purposes of encapsulation technique are (1) protection of core material from environmental conditions such as oxygen, temperature, moisture, RH, light; (2) masking of odor, taste, and activity of encapsulated materials; (3) controlled release of active compounds (sustained or delayed release); (4) separation of incompatible components;

(5) conversion of liquids to free-flowing solids; (6) increasing the oxidation stability and targeted release of encapsulated materials [14].

Microcapsule models can be classified into three basic categories as monocored, polycored, and matrix types (**Figure 2**) [14]. Monocored microcapsule contains a single hollow chamber within a capsule; however, the polycore microcapsule includes a number of different size chambers within the shell. On the hand, the matrix type of microparticle refers to active ingredients integrated within the matrix of the shell material [14]. Different types of microcapsules such as (i) simple microcapsule, (ii) matrix, (iii) irregular microcapsule, (iv) multicore microcapsule, (v) multiwall microcapsule, and (vi) assembly of microcapsule are shown in **Figure 3** [4].

It has been reported that the size and shape of microcapsules depend on wall materials and the methods used during preparation. The selection of wall materials relies on the properties of core materials, final products, and characteristics such as food-grade product, production cost, low viscosity property at high solid content, emulsifying properties and emulsion stability, ability of holding core materials in their structure without any reactivity during processing or storage, control release of core material, and protecting core materials from environmental conditions [15, 16]. The common wall materials of microencapsulated oil can be classified into three groups including carbohydrate, protein, and lipids and wax as summarized in **Table 1** [16].

An emulsion technique has been widely used for the preparation of encapsulation. Oil-in-water emulsions (O/W) are commonly used in cosmetics, pharmaceutical, and food industries for encapsulation by using different core materials.

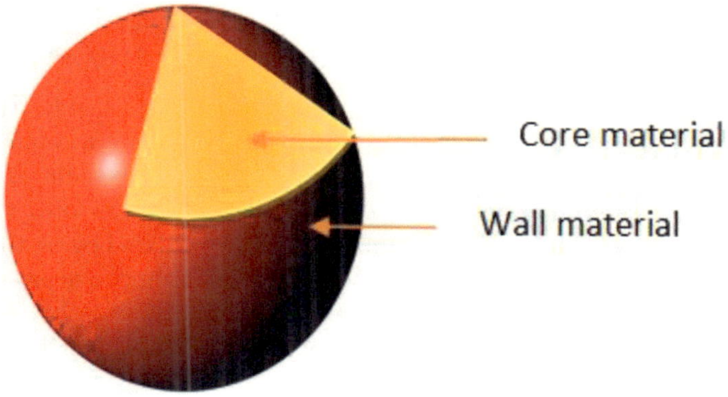

Figure 1.
Schematic diagram of microencapsulated material structure (adapted from [4]. Copyright 2015 by © Institute of Food Technologists).

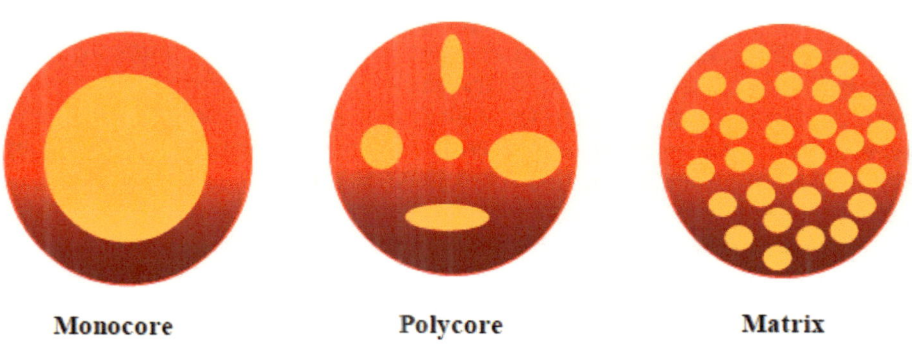

Figure 2.
Schematic diagram of microcapsules model (adapted from [14]. Copyright 2009 by DESIDOC.

Figure 3.
Different types of microcapsules: (i) simple microcapsule, (ii) matrix, (iii) irregular microcapsule, (iv) multicore microcapsule, (v) multiwall microcapsule, and (vi) assembly of microcapsule (adapted from [4]. Copyright 2015 by © Institute of Food Technologists).

Traditionally, O/W emulsions are prepared by oil homogenization with an aqueous phase containing one or more emulsifiers. However, the achievement of emulsion forming is limited depending on emulsifier properties such as on ionic strength, pH, and temperature, affecting emulsion stability and encapsulated compound [17, 18]. Guzey and McClements [17] indicated that one strategy to improve protection against environmental stresses is to create covalent protein-polysaccharide complexes and another strategy is to create multiple layers of emulsifiers and/or polyelectrolytes using a layer-by-layer (LBL) electrostatic deposition technique. According to LBL technique, it is based on LBL deposition of polyelectrolytes onto oppositely charged surfaces due to electrostatic attraction. Firstly, a primary emulsion containing an ionic emulsifier has produced a small oil droplet during homogenization. Thereafter, a secondary emulsion containing droplets coated with a two-layer interface is created using opposite charge polyelectrolytes with the primary emulsion. Finally, the secondary emulsion is mixed with another oppositely charge polyelectrolytes to create a tertiary emulsion. The procedure can be repeated to form oil droplet coated by interfaces containing more layer (**Figure 4**). The multilayer emulsions were reported having better stability to environmental stress than O/W emulsion with single-layer interfaces [3, 17, 19].

It has been found that LBL technique provides a multilayer emulsion with satisfying properties. However, the stable multilayer emulsions using an LBL technique depend on biopolymer properties, for example, charge density, molecular weight, conformation, emulsifier layer thickness, and bulk physicochemical condition. In addition, there have been several techniques applied for microencapsulation of vegetable oil powder. Drying process is the method commonly used for microencapsulation of vegetable oil, which changes liquid into powder. Spray drying is the most widely used

Carbohydrate	Proteins	Lipids and wax
Plant-based carbohydrate: • Maltodextrin • Starch • Cellulose • Gum arabic • Guar gum • Pectin • Galactomannans • Cyclodextrin • Mesquite gum etc. Marine-based carbohydrate: • Carrageenan • Alginate Microbial- or animal-based carbohydrate: • Xanthan • Chitosan • Dextran • Gellan	Plant-based protein: • Soy protein • Pea protein • Barley protein • Zein • Gluten Animal-based protein: • Casein • Whey protein • Gelatin	• Milk fat • Phospholipid • Beeswax • Carnauba wax

Source: [16].

Table 1.
Different wall materials used for microencapsulation.

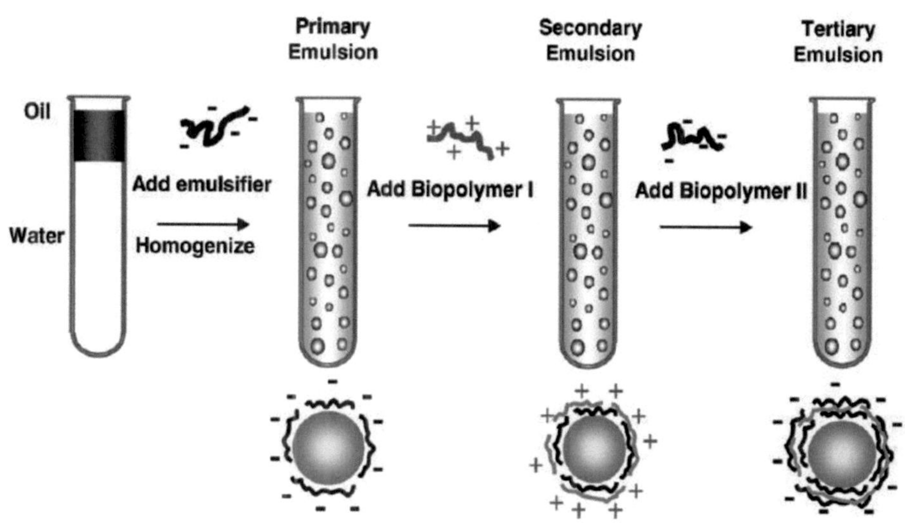

Figure 4.
Schematic representation of layer-by-layer technique producing multilayer emulsions (reproduced with permission from [17].

encapsulation technique in the food industry that is a relatively simple, continuous, and low-cost commercial process [4]. The microencapsulation using spray drying involves atomization and drying of solution, emulsion, suspension, slurry, and paste to produce solid material. It contains (1) preparation of emulsion sample, (2) atomization of the emulsion into fine droplets, (3) droplet-hot-air contact, (4) evaporation of droplet water, and (5) recovery of powder (**Figure 5**) [20]. Generally, the spray drying

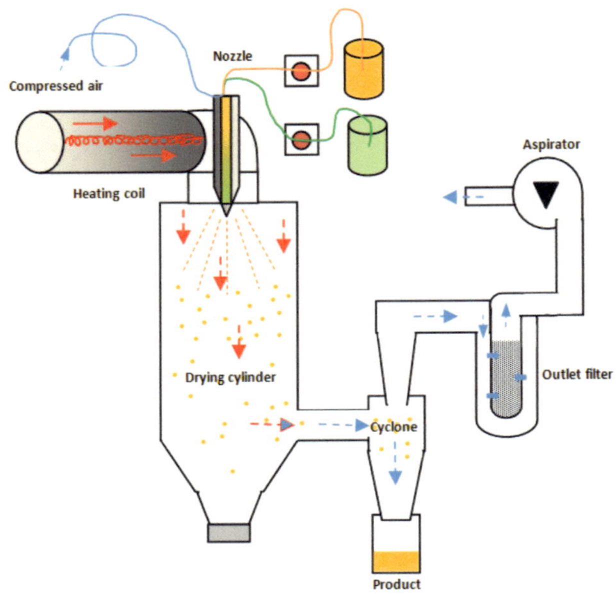

Figure 5.
Schematic representation of spray dryer.

has been used to produce the encapsulation of vegetable oil in the food industry [8, 9, 21–24]. The spray drying process conditions (inlet and outlet temperature, nozzle size, feed rate, etc.) have been found to affect the characteristics and properties of encapsulations. However, the optimum drying condition should obtain minimized fat-free surface powder. It was reported that low inlet and outlet temperatures can reduce the viscosity and the diffusivity of fat. Moreover, large emulsion droplet and nozzle size provide a large powder with low surface area and low fat-free surface [25–28]. The advantages of spray drying compose of simple process, fast and easy to scale up, availability of machinery, low production cost, varied particle sizes, and excellent dispersibility in media. However, some limitations of spray drying were stated such as loss of core material during processing and oxidation of flavoring compounds [29, 30]. In addition, not only spray drying technique was selected to apply for encapsulation process, but different drying techniques are also available for vegetable oil encapsulation such as freeze drying, fluidized bed spray drying, nozzleless electrostatic atomization spray drying, and supercritical carbon dioxide spray drying [25, 26, 31, 32].

3. Vegetable oil-in-water emulsion preparation using microencapsulation technique

Vegetable oils such as soybean, sesame, sunflower, flaxseed, coconut, rice, etc. have been found as major sources of edible oil, which takes place for almost 70% of edible oil. They are composed primarily of triglycerides, esters of one molecule of glycerol, and three molecules of fatty acids. Triglycerides are the most abundant component found in lipid (more than 90%). On the other hand, a vague variation of free fatty acid was reported (0.3–9.5%) depending on a plant source [33]. Different fatty acids were classified by the nature of the hydrocarbon chain. This chain length can vary from 4 to 24 carbon atoms and can be classified as saturated (without a double bond, SFA), monounsaturated (one double bond, MUFA), or polyunsaturated (two or more double bonds, PUFA), which contains 18 carbons with different saturation degree. The position of fatty acids on the glycerol molecule can be in the 1, 3, or 2 positions depending on saturation. Glycerides with saturated fatty acid in this position usually have a high melting point and poor solubility, which can cause nutritional problems and poor digestibility [33].

Generally, fatty acid composition of vegetable oils is formed by a mixture of saturated (SFAs) and unsaturated (UNFAs) fatty acids classified by a number of unsaturated bonds as monounsaturated (MUFAs) or polyunsaturated fatty acids (PUFAs). Nevertheless, each of the analyzed vegetable oils has specific fatty acid distribution depending on their plant sources. Thus, their impact on human health could be assessed according to individual fatty acids because of their different influences on human health and risks of serious diseases [31]. Recently, nutritionists have recommended vegetable oils as an important part of a healthy diet due to their high contents of fatty acids (FAs) [31].

3.1 Soybean oil

Soybean oil is the important seed oil produced in the world due to its high quality and low-cost production. Soybean oil has 15% of saturated fatty acid and 80.7% of unsaturated fatty acid (50.8% linoleic acid content and 6.8% linolenic acid). Soybean oil contains a saturated, monounsaturated, and polyunsaturated fats in healthy proportions (SFA:MUFA:PUFA = 16:24:58). Linoleic acid (omega-6) is the major polyunsaturated fatty acid found in oil; phytosterols, especially B-sitosterol, inhibit cholesterol absorption and reduce blood LDL cholesterol levels by 10–15%. Moreover, it contains several bioactive compounds such as antioxidant vitamin E, a powerful lipid soluble vitamin, which is important to maintain the integrity of cell membranes and protect them from harmful reactive oxygen-free radicals; vitamin K, an essential element in promoting bone formation and strengthening, and neuronal protection in the brain [32]. Soybean oil provides several advantages and disadvantages compared to other vegetable oils. The advantages of soybean oil include (1) a high content of unsaturated fatty acid, (2) wide temperature range of liquid state, (3) hydrogenated selectively for blending with semisolid and liquid oil, (4) source of nutrients (such as flavonoids and isoflavonoids, phenolic acids, phytoalexins, phytosterols, proteins, and peptides) and mineral (such as copper, manganese, molybdenum, phosphorus, potassium, B vitamin, and omega-3 fatty acids (alpha-linolenic acid) [32, 33]. The disadvantages of soybean oil display a large number of phosphatides, which have to be removed by processing technique and high levels of linolenic acid, which is responsible for its flavor and odor reversion [33]. Several encapsulation techniques for soybean oil have been studied [3, 34]. Spray-dried soybean oil emulsions were made of whey protein, lactose, and soybean oil. It is reported that the ability of whey protein used for soybean oil encapsulation is moderate and the soybean was found in low quality when compared to sodium caseinate under dry and humid atmosphere storage condition (RH 75% for 4 days). Moreover, the fat releasing was observed on powder surface due to some critical amount of lactose containing in powder comparing the powder containing a small amount of lactose [21]. Moreover, the application of electrostatic atomization for soybean oil encapsulation has been reported [25]. W/O emulsion containing glycine and taurine as the wall materials was prepared. The result was shown that the oxidative stability of soybean oil during high-temperature storage was improved.

3.2 Sesame oil

Sesame oil contains 80% unsaturated fatty acids. Oleic and linoleic acids are the main fatty acids. It contains less than 20% of a saturated fatty acid including palmitic acid and stearic acid. The fatty acid composition is 40.7–49.3% of linoleic,

29.3–41.4% oleic, 8.0–10.3% palmitic, and 2.1–4.8% stearic acids in the seed oil [35]. Moreover, sesame oil is also rich in γ-tocopherols (90.5%) [36]. The crude sesame oil contains lignans such as sesamin (293–885 mg/100 g oil), sesamolin (123–459 mg/100 g oil), and sesamol (trace–5.6 mg/100 g oil) [33]. The sesame lignans have been reported to inhibit lipid oxidation and to enhance antioxidant activity of vitamin E in lipid peroxidation systems [37]. All these lignans have multiple physiological functions including inhibiting cholesterol absorption from the intestine, reducing 3-hydroxy-3-methyl-glutaryl CoA reductase activity in the liver microsomes [38], and inhibiting hepatic endoplasmic reticulum stress and apoptosis in high-fat-diet-fed mice [39]. Onsaard et al. [40] applied a multilayer emulsion for sesame oil aiming to investigate the influence of maltodextrin and environmental stresses (pH, NaCl, and sucrose) on the stability of sesame oil-in-water emulsions containing droplets stabilized by WPC-k-carrageenan membranes. The primary emulsion containing whey protein concentrate-coated droplets was prepared by homogenization. The secondary emulsion containing whey protein concentrate-k-carrageenan was produced by mixing of the primary emulsion with an aqueous k-carrageenan in the absence or presence of maltodextrin solution. There were no significant changes in mean droplet diameter and z-potential of droplets at any maltodextrin concentration (0–30%) or a dextrose equivalent (10 and 15) after 24 h storage. The apparent viscosity of emulsions was increased when the maltodextrin concentration increased. The secondary emulsion containing 15% maltodextrin with dextrose equivalent of 10 provided the stability to aggregate at pH 6–8, NaCl 300 mM, and sucrose 0–20% [40]. Onsaard et al. [8] also studied the oxidation stability of encapsulated sesame oil powder by spray drying. Microencapsulated sesame oil powder was prepared from sesame oil-in-water emulsions containing 15% sesame oil, 0.5% whey protein concentrate, 0.2% κ-carrageenan, and 0–30% maltodextrin with a dextrose equivalent (DE) of 10 using spray drying method. They found that the microencapsulated powder provided high encapsulation yields (86.73%) and low moisture content (3.19%) and water activity (aw = 0.28). The powder exhibited a spherical shape with a few cracks on the surface. They reported that no significant difference in TBARS value was observed during storage at ambient temperature, cold storage temperature, and frozen temperature for 30 days storage ($p > 0.05$). They also suggested that using κ-carrageenan as a secondary layer can improve oxidation stability. They suggested that to the emulsion containing anionic droplets stabilized by interfacial membranes, comprising whey protein concentrate/k-carrageenan/maltodextrin can be used to produce microencapsulation of sesame oil using spray drying technique. Therefore, the powder performed better in protecting the sesame oil against oxidation during storage [8].

3.3 Sunflower oil

Sunflower oil contains a high content of polyunsaturated fatty acids (PUFA) mainly linoleic acid (18:2 n-6) including 68–72% of total fatty acid content. Moreover, it is considered to display an excellent hypocholesterolemic action, which can reduce cardiovascular risk [41]. The other important component of sunflower oil is vitamin E (α-tocopherol). Its high level of vitamin E is helpful for antioxidant activity [42]. Sunflower oil has been encapsulated in starch matrices (native potato starch, water, glycerol, and emulsifier) by extrusion. Extrusion processing parameters such as screw speed, the presence of die head, throughput, melt temperature, and especially the screw configuration play an important role in the development of the dispersed phase morphology [43]. Domian et al. [44] studied sunflower oil microencapsulated using a spray drying method in the matrix of trehalose and whey protein isolate or sodium caseinate. The microencapsulated powder was able to prevent oil oxidation after observing agglomeration during 3 months storage [44]. Belingheri et al. [45] reported that high-oleic sunflower oil carried on porous starch as an alternative to spray drying does not undergo significantly higher oxidation than traditionally spray-dried sunflower oil. They have suggested that plating on porous starch could be a suitable technological alternative to spray drying for flavor encapsulation [45].

3.4 Flaxseed oil

Flaxseed oil is a great source of ω-3 fatty acids. It contains 73% polyunsaturated fatty acids (PUFA), 9% saturated fatty acids, and 18% monosaturated fatty acids [46]. Major fatty acids in flaxseed oil are α-linolenic acid (c18:3; ω-3) (39.90–60.42%), linoleic acid (c18:2; ω-6) (12.25–17.44%), oleic acid (c18:1) (13.44–19.39%), stearic acid (c18:0) (2.24–4.59%), and palmitic acid (c16:0) (4.90–8.00%) [47]. α-Linolenic acid is an essential fatty acid as a precursor of the important long-chain polyunsaturated fatty acid eicosapentaenoic (EPA) and docosahexaenoic acid (DHA) [48]. Goyal et al. (2014) reported that flaxseed oil, fibers, and flax lignans benefit to the reduction of cardiovascular disease, atherosclerosis, diabetes, cancer, arthritis, osteoporosis, autoimmune, and neurological disorders [49]. Although the flaxseed oil is high in antioxidant activity, it can be oxidized after extraction and purification. Microencapsulation technology was suggested to protect PUFAs oil against oxidation, improving their manipulation, modulating their release, and masking their unpleasant test and odor. Increasing the stability of flaxseed oil by microencapsulation process is based on ionic gelation through vibrating nozzle extrusion technology, using pectin as wall material [48]. The authors applied two different drying methods, passive air drying, and fluidized bed drying. The results show that the fluidized bed drying method provided the 20-fold faster and higher payload. Under accelerated storage, higher stability of the encapsulated flaxseed oil powder was found compared to bulk oil [48]. Rubilar et al. [50] optimized the process condition to improve the microencapsulation efficiency of flaxseed oil using a spray drying technique. The results showed that higher microencapsulation efficiency values were obtained with a high concentration of encapsulating wall (30% wall material concentration, 14% oil concentration, and maltodextrin/gum arabic wall type). The microencapsulation of flaxseed oil can enhance the oxidation stability, which can be applied for soup powder enriched with microencapsulated flaxseed oil as a source of ω-3 [50]. Spray-dried flaxseed oil emulsions were prepared by chickpea or lentil protein isolate and maltodextrin. The oxidation stability of encapsulated flaxseed oil was found over a storage period of 25 days at room temperature, and 84.2% of the encapsulated flaxseed oil within the gastrointestinal environments was delivered [51].

3.5 Coconut oil

Coconut oil is edible oil extracting from a kernel of mature coconut palm (*Cocos nucifera*). The coconut oil is the white or slightly yellowish color at a temperature above 26°C and its strong odor or flavor is due to δ- and γ-lactones [2]. The oil contains triacylglycerols (84.0–93.1%), 1,2-diacylglycerols (1.5–5.1%), 1,3-diacylglycerols (1.2–2.1%), monoglyceride (1.0–7.0%), free fatty acids (1.0–2.6%), phospholipids (0.03–0.4%), and glycolipids (0.2–0.35%) [52]. Hui et al. [33] have reported that coconut oil contains 90% saturated fatty acids and 10% unsaturated fatty acids. Medium chain triglycerides (MCTs) are the main components of a fatty acid containing lauric acid (40–50%), myristic acid (13–19%), and

Encapsulating ingredient	Wall material	Encapsulation process	Encapsulation Efficiency (EE) Encapsulation yield (EY)	Oil content	Particle size	References
Soybean oil	Whey protein/lactose	Spray drying	—	30%	0.4 μm	[25]
	Taurine and glycine	Nozzleless electrostatic atomization	—	2.35% and 8.56%	~0–15 μm	
Sesame oil	Whey protein concentrate, κ-carrageenan, and maltodextrin	Spray drying	EY 86.73%	15%	570–650 nm	[8, 40]
Sunflower oil	Native potato starch/glycerol/emulsifier	Extrusion	—	4 μl	15.3–53.4 μm	[43]
	Trehalose/whey protein isolate or sodium caseinate (NaCas)	Spray drying	EE 96–99%	22%	10–70 μm	[44]
	Maltodextrin / hydroxypropylmethylcellulose	Spray drying	EE 73.13–87.00%	—	—	[9]
	Gum arabic and maltodextrin/porous starch	Spray drying	—	20%	—	[45]
Flaxseed oil	Pectin	Vibrating nozzle extrusion/ fluid bed	EE 98%	15%	862–1463 μm	[48]
	Maltodextrin/gum arabic	Spray drying	EE 54.6–90.7%	14 and 20%	17.6 and 23.1 μm	[50]
	Chickpea protein/lentil protein isolate/ maltodextrin	Spray drying	EE 88 and 86.3%	20%	16.3–24.0 μm 21.0–26.1 μm	[51]
Coconut oil	Gelatin solution and maltodextrin	Spray drying	EE ~82% EY ~90%	14.66%	10.3–6.0 μm	[22]
	Maltodextrin/sodium caseinate/soy lecithin	Supercritical carbon dioxide spray drying	EE 73–80%	11.6%	27–72 μm	[57]
Rice bran oil	Tapioca starch/soya protein isolate	Spray drying	EE 76.97%	20%	—	[62]
	Jackfruit seed starch/whey protein isolate	Spray drying	EE 85.90%	20%	3.40–300.51μm	[63]

Table 2.
Application of encapsulated vegetable oils using different drying techniques.

palmitic acid (4–18%) [33]. Coconut oil, especially virgin coconut oil (VCO), has been claimed as a health benefit product such as antioxidant, anti-inflammatory, lipid-lowering, and cytoprotective efficacies due to its higher polyphenolics [53]. It also has been reported that coconut oil exhibited antioxidant property and prevented the peroxidation of lipids both in vitro and in vivo conditions [54]. Moreover, MCTs have been reported as a human health benefit such as weight and glucose control, as well as lipid metabolism and acting as a tumor inhibitor when consumed in a diet [55, 56]. Application of ultrasound for microencapsulation of coconut milk fat using spray drying has been studied by Le et al. [22]. It was reported that using a mixture of coconut milk, gelatin solution, and maltodextrin as a wall material was found successful [22]. On the other hand, VCO microcapsules from oil-in-water (O/W) emulsion using supercritical carbon dioxide spray drying have been reported by Hee et al. [57]. The authors prepared an O/W emulsion by using maltodextrin, sodium caseinate, and soy lecithin as wall materials before supercritical carbon dioxide spray drying was conducted. This result has found a minor effect on antioxidant activity and fatty acids composition of encapsulated coconut oil [57].

3.6 Rice bran oil

Rice bran oil can be extracted from a hard outer brown layer of rice caryopsis during milling. The compositions of curd rice bran oil are 90–96% saponifiable lipids, 83–96% triacylglycerols (TAG), 3–4% diglyceride, 6–7% monoglyceride, 2–4% free fatty acids, 3–4% waxes, 6–7% glycolipids, 4–5% phospholipids, and 4.2% unsaponifiable lipids. The fatty acid compositions of rice bran oil are 0.3% myristic acid (C14:0), 15.0% palmitic acid (C16:0), 1.7% stearic acid (C18:0), 43.0% oleic acid (C18:1), 37.4% linoleic acid (C18:2; ω-6), and 1.5% linolenic acid (C18:3; ω-3) [33]. Rice bran oil contains vitamin E (α-tocopherol, β-tocopherol, α-tocotrienol, and β-tocotrienol), γ-oryzanol, and phytosterols, which are known as antioxidants [58, 59]. In addition, rice bran oil provides several health benefits such as reducing cholesterol, cardiovascular health benefits, and antitumor activity [60, 61]. Microencapsulation of rice bran oil has been reported using a spray drying technique at 140°C inlet air temperature and a combination of different wall materials (tapioca starch and soy protein isolate). This encapsulated rice bran oil powder provided a high encapsulation efficiency and high γ-oryzanol content with low peroxide value [62]. Moreover, Murali et al. [63] optimized rice bran oil encapsulation condition using jackfruit seed starch and whey protein isolate blend as wall materials by spray drying technique. They found that rice bran oil emulsion made with 20% rice bran oil, 3:1 of jackfruit seed starch and whey protein isolate ratio, and 140°C spray drying inlet temperature provided a high encapsulation efficiency and low peroxide value microcapsules [63].

According to several researches reported and reviewed in this chapter, the application of encapsulated vegetable oil offers several benefits to the food industry are summarized in **Table 2**.

4. Characterization of microencapsulated vegetable oil

There are several encapsulated vegetable oil powder characteristics used for characterization of encapsulated powder aiming to ensure that encapsulation techniques can be applied to stabilize the vegetable oil powder in physical, chemical, and physicochemical properties as concluded in **Table 3**.

Encapsulated characterizations	Indicators	Measurements	References
Particle characterizations	Particle size	Dynamic light scattering technique	[8, 44, 57, 66]
	Distribution and mean particle size	Laser light diffraction	[8, 44, 57]
	Zeta potential (ζ-potential)	Surface charge	[8]
	Particle morphology	Scanning electron microscopy (SEM) or transmission electron microscopy (TEM)	[8, 21, 23, 25, 48, 57, 65]
	Moisture content	Hot air oven moisture analyzer	[23, 57, 66]
	Bulk density	Volumeter	[44]
Oxidative stability (under accelerated storage conditions)	Peroxide value	Peroxide	[8, 44, 48]
	2-Thiobarbituric acid reactive substances (TBARS)	Malondialdehyde	[8]
	p-Anisidine value (p-AV)	p-Anisidine	[48]
Thermal analysis	Melting point Thermal profile	Differential scanning calorimeter (DSC) Thermal gravimetric analysis Dynamic mechanical analysis	[67]
Amount or payload	Oil content	High-performance liquid chromatography (HPLC), gas chromatography (GC), or gas chromatography/mass spectrometry (GC/MS)	[25, 48, 57, 66]
	Interaction between materials in encapsulated emulsions	Fourier transform infrared (FT-IR) technique	[23]
	The difference between total and free oil concentration on encapsulation	Encapsulation efficiency (EE)	[57, 66]

Table 3.
Different characterization of microencapsulated vegetable oils.

4.1 Particles characterization

- Particle size, distribution, and mean particle size can be determined by dynamic light scattering technique that has the advantage of being fast and noninvasive, but it does require low particle concentrations [64].

- Zeta potential (ζ-potential) of particles is a scientific term for electrokinetic potential in particles dispersions and a measure of surface charge of particles that reflects their long-term stability.

- Particle morphology: scanning electron microscopy (SEM), transmission electron microscopy (TEM), and electron spectroscopy can visualize surface morphology, dispersed and agglomerated particles, and surface functionalization [65]; in addition, transmission electron microscopy (TEM) or dynamic light scattering.

- Moisture content and bulk density of encapsulated powders.

4.2 Oxidative stability of encapsulated particles

In an approach to evaluating the oxidative stability of oil, the encapsulated particles are evaluated for oxidation at storage times, for example, peroxide value,

2-thiobarbituric acid reactive substances (TBARS) and headspace analysis used to determine the production of propanal and hexanol as indicators of vegetable oil oxidation.

4.3 Thermal analysis

This is one of the popular analyses where the properties of microencapsulated vegetable oil are studied as they change with temperature. Several methods are used:

- Differential scanning calorimeter (DSC): heat flow changes versus temperature or time
- Thermal gravimetric analysis: mass change versus temperature or time
- Dynamic mechanical analysis: measures storage modulus (stiffness) and loss modulus (damping) versus temperature, time, and frequency

4.4 Content or payload

- High-performance liquid chromatography (HPLC)
- Gas chromatography (GC) or gas chromatography/mass spectrometry (GC/MS)
- Fluorescent
- Fourier transform infrared (FT-IR) technique
- Thermal gravimetric analysis
- Encapsulation efficiency (EE)

5. Conclusion

Vegetable oil can be stabilized by using microencapsulated technique. The successful preparation technique is suggested by using multilayer emulsion followed by spray drying. Mostly, wall materials are prepared from a mixture of polysaccharide and protein in order to coat vegetable oil droplets by means of creating covalent protein-polysaccharide bond with a slight usage of emulsifier. The encapsulated vegetable oil powder expresses low oxidative deterioration with chemical and thermal stabilities, which can be applied in different food systems.

Acknowledgements

The authors are grateful to our undergraduate and postgraduate students who give the best of their laboratory works as well as the Indigenous Food Research and Industrial Development Unit, Faculty of Agriculture, Ubon Ratchathani University for all the facilities. The best special sincere appreciation is given to distinguished Professor David Julian McClements for providing the inspiration to the authors in the field of food biopolymers and colloids.

Conflict of interest

The authors declare no conflict of interest.

Author details

Ekasit Onsaard* and Wiriya Onsaard
Indigenous Food Research and Industrial Development Unit, Department of Agro-Industry, Faculty of Agriculture, Ubon Ratchathani University, Ubon Ratchathani, Thailand

*Address all correspondence to: ekasit.o@ubu.ac.th

References

[1] USDA. Oilseeds: World Markets and Trade; 2018

[2] Salunkhe DK. World Oilseeds: Chemistry, Technology, and Utilization. New York: Van Nostrand Reinhold; 1992

[3] Aluyor EO, Ori-Jesu M. The use of antioxidants in vegetable oils—A review. African Journal of Biotechnology. 2008;7(25):4836-4842

[4] Bakry AM, Abbas S, Ali B, Majeed H, Abouelwafa MY, Mousa A, et al. Microencapsulation of oils: A comprehensive review of benefits, techniques, and applications. Comprehensive Reviews in Food Science and Food Safety. 2016;**15**(1):143-182

[5] Goyal A, Sharma V, Sihag MK, Tomar SK, Arora S, Sabikhi L, et al. Development and physico-chemical characterization of microencapsulated flaxseed oil powder: A functional ingredient for omega-3 fortification. Powder Technology. 2015;**286**:527-537

[6] Gupta PK, Jadhav SB, Singhal RS. Development of shrikhand premix using microencapsulated rice bran oil as fat alternative and hydrocolloids as texture modifier. Food Hydrocolloids. 2015;**48**:220-227

[7] Ifeduba EA, Akoh CC. Microencapsulation of stearidonic acid soybean oil in Maillard reaction-modified complex coacervates. Food Chemistry. 2016;**199**:524-532

[8] Onsaard E, Putthanimon J, Singthong J, Thammarutwasik P. Oxidation stability of sesame oil encapsulated by spray drying. International Food Research Journal. 2018;**25**(2):784-792

[9] Roccia P, Martínez ML, Llabot JM, Ribotta PD. Influence of spray-drying operating conditions on sunflower oil powder qualities. Powder Technology. 2014;**254**:307-313

[10] Rojas VM, Marconi LFCB, Guimarães-Inácio A, Leimann FV, Tanamati A, Gozzo ÂM, et al. Formulation of mayonnaises containing PUFAs by the addition of microencapsulated chia seeds, pumpkin seeds and baru oils. Food Chemistry. 2019;**274**:220-227

[11] Gharsallaoui A, Roudaut G, Chambin O, Voilley A, Saurel R. Applications of spray-drying in microencapsulation of food ingredients: An overview. Food Research International. 2007;**40**(9):1107-1121

[12] Shahidi F, Han XQ. Encapsulation of food ingredients. Critical Reviews in Food Science and Nutrition. 1993;**33**(6):501-547

[13] Singh MN, Hemant KSY, Ram M, Shivakumar HG. Microencapsulation: A promising technique for controlled drug delivery. Research in Pharmaceutical Sciences. 2010;**5**(2):65-77

[14] Dubey R, Shami TC, Bhasker Rao KU. Microencapsulation technology and applications. Defence Science Journal. 2009;**59**(1):14

[15] Desai KGH, Jin Park H. Recent developments in microencapsulation of food ingredients. Drying Technology. 2005;**23**(7):1361-1394

[16] Özbek ZA, Ergönül PG. A review on encapsulation of oils. Celal Bayar University Journal of Science. 2017;**3**(2):293-309

[17] Guzey D, McClements DJ. Formation, stability and properties of multilayer emulsions for application in the food industry. Advances in Colloid and Interface Science. 2006;**128-130**:227-248

[18] Burgos-Díaz C, Wandersleben T, Marqués AM, Rubilar M. Multilayer emulsions stabilized by vegetable proteins and polysaccharides. Current Opinion in Colloid and Interface Science. 2016;**25**:51-57

[19] Ogawa S, Decker EA, McClements DJ. Influence of environmental conditions on the stability of oil in water emulsions containing droplets stabilized by lecithin–chitosan membranes. Journal of Agricultural and Food Chemistry. 2003;**51**(18):5522-5527

[20] Haque MA, Timilsena Y, Adhikari B. Book Chapter Spray Drying; 2015

[21] Fäldt P, Bergenståhl B. Spray-dried whey protein/lactose/soybean oil emulsions. 1. Surface composition and particle structure. Food Hydrocolloids. 1996;**10**(4):421-429

[22] Le HD, Le VV. Application of ultrasound to microencapsulation of coconut milk fat by spray drying method. Journal of Food Science and Technology. 2015;**52**(4):2474-2478

[23] Matsuura T, Ogawa A, Tomabechi M, Matsushita R, Gohtani S, Neoh TL, et al. Effect of dextrose equivalent of maltodextrin on the stability of emulsified coconut-oil in spray-dried powder. Journal of Food Engineering. 2015;**163**:54-59

[24] Fioramonti SA, Stepanic EM, Tibaldo AM, Pavón YL, Santiago LG. Spray dried flaxseed oil powdered microcapsules obtained using milk whey proteins-alginate double layer emulsions. Food Research International. 2018;**119**:931-940

[25] Mori C, Kadota K, Tozuka Y, Shimosaka A, Yoshida M, Shirakawa Y. Application of nozzleless electrostatic atomization to encapsulate soybean oil with solid substances. Journal of Food Engineering. 2019;**246**:25-32

[26] Turchiuli C, Fuchs M, Bohin M, Cuvelier ME, Ordonnaud C, Peyrat-Maillard MN, et al. Oil encapsulation by spray drying and fluidised bed agglomeration. Innovative Food Science & Emerging Technologies. 2005;**6**(1):29-35

[27] Pisecky J. Handbook of Milk Powder Manufacture. Søborg: Niro; 1997

[28] Keogh, M.K. Spray dried microencapsulated fat powders. In: Onwulata C, editor. Encapsulated and Powdered Foods. New York: CRC Press; 2005. pp. 477-484

[29] Jackson LS, Lee K. Microencapsulation and the Food Industry. Lebensmittel Wissenschaft und Technologie.1991;**24**:289-297

[30] Jeyakumari A, Zynudheen AA, Parvathy U. Microencapsulation of bioactive food ingredients and controlled release—A review. MOJ Food Processing & Technology. 2016;**2**(6):214-224

[31] Orsavova J, Misurcova L, Ambrozova JV, Vicha R, Mlcek J. Fatty acids composition of vegetable oils and its contribution to dietary energy intake and dependence of cardiovascular mortality on dietary intake of fatty acids. International Journal of Molecular Sciences. 2015;**16**:12871-12890

[32] Varnham A. Seed Oil: Biological Properties, Health Benefits and Commercial Applications. UK: Nova Science Publishers; 2014

[33] Hui YH. Bailey's industrial oil and fat products. In: Edible Oil and Fat Products : Oils and Oilseeds. Vol. 2. New York/Chichester: Wiely-Interscience Publication/John Wiley & Sons, Inc; 1996

[34] Chatterjee D, Bhattacharjee P. Comparative evaluation of the antioxidant efficacy of encapsulated

and un-encapsulated eugenol-rich clove extracts in soybean oil: Shelf-life and frying stability of soybean oil. Journal of Food Engineering. 2013;**117**(4):545-550

[35] Uzun B, Arslan Ç, Furat Ş. Variation in fatty acid compositions, oil content and oil yield in a germplasm collection of sesame (*Sesamum indicum* L.). Journal of the American Oil Chemists Society. 2008;**85**(12):1135-1142

[36] Gharby S, Harhar H, Bouzoubaa Z, Asdadi A, El Yadini A, Charrouf Z. Chemical characterization and oxidative stability of seeds and oil of sesame grown in Morocco. Journal of the Saudi Society of Agricultural Sciences. 2017;**16**(2):105-111

[37] Ghafoorunissa, Hemalatha S, Rao MV. Sesame lignans enhance antioxidant activity of vitamin E in lipid peroxidation systems. Molecular and Cellular Biochemistry. 2004;**262**(1):195-202

[38] Hirose N, Inoue T, Nishihara K, Sugano M, Akimoto K, Shimizu S, et al. Inhibition of cholesterol absorption and synthesis in rats by sesamin. Journal of Lipid Research. 1991;**32**(4):629-638

[39] Kim M, Woo M, Noh JS, Choe E, Song YO. Sesame oil lignans inhibit hepatic endoplasmic reticulum stress and apoptosis in high-fat diet-fed mice. Journal of Functional Foods. 2017;**37**:658-665

[40] Onsaard E, Putthanimon J, Singthong J, Thammarutwasik P. Influence of maltodextrin and environmental stresses on stability of whey protein concentrate/κ-carrageenan stabilized sesame oil-in-water emulsions. Food Science and Technology International. 2014;**20**(8):617-628

[41] Upadhyay R, Mishra HN. Predictive modeling for shelf life estimation of sunflower oil blended with oleoresin rosemary (*Rosmarinus officinalis* L.) and ascorbyl palmitate at low and high temperatures. LWT-Food Science and Technology. 2015;**60**(1):42-49

[42] Muhammad Anjum F, Nadeem M, Issa Khan M, Hussain S. Nutritional and therapeutic potential of sunflower seeds: A review. British Food Journal. 2012;**114**(4):544-552

[43] Yılmaz G, Jongboom ROJ, Feil H, Hennink WE. Encapsulation of sunflower oil in starch matrices via extrusion: Effect of the interfacial properties and processing conditions on the formation of dispersed phase morphologies. Carbohydrate Polymers. 2001;**45**(4):403-410

[44] Domian E, Sułek A, Cenkier J, Kerschke A. Influence of agglomeration on physical characteristics and oxidative stability of spray-dried oil powder with milk protein and trehalose wall material. Journal of Food Engineering. 2014;**125**:34-43

[45] Belingheri C, Giussani B, Rodriguez-Estrada MT, Ferrillo A, Vittadini E. Oxidative stability of high-oleic sunflower oil in a porous starch carrier. Food Chemistry. 2015;**166**:346-351

[46] Cunnane SC, Ganguli S, Menard C, Liede AC, Hamadeh MJ, Chen Z-Y, et al. High α-linolenic acid flaxseed (*Linum usitatissimum*): Some nutritional properties in humans. The British Journal of Nutrition. 1993;**69**(2):443-453

[47] Goyal A, Sharma V, Upadhyay N, Gill S, Sihag M. Flax and flaxseed oil: An ancient medicine & modern functional food. Journal of Food Science and Technology. 2014;**51**(9):1633-1653

[48] Menin A, Zanoni F, Vakarelova M, Chignola R, Donà G, Rizzi C, et al. Effects of microencapsulation by ionic gelation on the oxidative stability

of flaxseed oil. Food Chemistry. 2018;**269**:293-299

[49] Holstun J, Zetocha D. An Analysis of Flaxseed Utilization in the Health Food Industry: Institute for Business and Industry Development. Fargo: North Dakota State University; 1994

[50] Rubilar M, Morales E, Contreras K, Ceballos C, Acevedo F, Villarroel M, et al. Development of a soup powder enriched with microencapsulated linseed oil as a source of omega-3 fatty acids. European Journal of Lipid Science and Technology. 2012;**114**(4):423-433

[51] Can Karaca A, Low N, Nickerson M. Encapsulation of flaxseed oil using a benchtop spray dryer for legume protein-maltodextrin microcapsule preparation. Journal of Agricultural and Food Chemistry. 2013;**61**(21):5148-5155

[52] Gopalakrishnan N, Narayanan CS, Mathew AG, Arumughan C. Lipid composition of coconut cake oil. Journal of the American Oil Chemists Society. 1987;**64**(4):539-541

[53] Narayanankutty A, Illam SP, Raghavamenon AC. Health impacts of different edible oils prepared from coconut (*Cocos nucifera*): A comprehensive review. Trends in Food Science & Technology. 2018;**80**:1-7

[54] Nevin KG, Rajamohan T. Virgin coconut oil supplemented diet increases the antioxidant status in rats. Food Chemistry. 2006;**99**(2):260-266

[55] Cohen LA, Thompson DO. The influence of dietary medium chain triglycerides on rat mammary tumor development. Lipids. 1987;**22**(6):455-461

[56] Marten B, Pfeuffer M, Schrezenmeir J. Medium-chain triglycerides. International Dairy Journal. 2006;**16**(11):1374-1382

[57] Hee YY, Tan CP, Rahman RA, Noranizan M, Smith RL, Chong GH. Production of virgin coconut oil microcapsules from oil-in-water emulsion with supercritical carbon dioxide spray drying. The Journal of Supercritical Fluids. 2017;**130**:118-124

[58] Patel M, Naik SN. Gamma-oryzanol from rice bran oil—A review. Journal of Scientific & Industrial Research. 2004;**63**:569-578

[59] Nagendra Prasad MN, Sanjay KR, Shravya Khatokar M, Vismaya MN, Nanjunda SS. Health benefits of rice bran—A review. Journal of Nutrition & Food Sciences. 2011;**1**(3):1-7

[60] Hu W, Wells JH, Shin T-S, Godber JS. Comparison of isopropanol and hexane for extraction of vitamin E and oryzanols from stabilized rice bran. Journal of the American Oil Chemists Society. 1996;**73**(12):1653-1656

[61] Lichtenstein AH, Ausman LM, Carrasco W, Gualtieri LJ, Jenner JL, Ordovas JM, et al. Rice bran oil consumption and plasma lipid levels in moderately hypercholesterolemic humans. Arteriosclerosis and Thrombosis. 1994;**14**(4):549-556

[62] Murali S, Kar A, Patel AS, Kumar J, Mohapatra D, Dash SK. Encapsulation of rice bran oil in tapioca starch-soya protein isolate complex using spray drying. Indian Journal of Agricultural Sciences. 2016;**86**(8):984-991

[63] Murali S, Kar A, Patel Avinash S, Mohapatra D, Krishnakumar P. Optimization of Rice bran oil encapsulation using jackfruit seed starch—whey protein isolate blend as wall materialand its characterization. International Journal of Food Engineering. 2017;**13**(4)

[64] Ross Hallett F. Particle size analysis by dynamic light scattering.

Food Research International. 1994;**27**(2):195-198

[65] Singh AK. Chapter 4—Experimental methodologies for the characterization of nanoparticles. In: Singh AK, editor. Engineered Nanoparticles. Boston: Academic Press; 2016. pp. 125-170

[66] Gallardo G, Guida L, Martinez V, López MC, Bernhardt D, Blasco R, et al. Microencapsulation of linseed oil by spray drying for functional food application. Food Research International. 2013;**52**(2):473-482

[67] Singh VK, Banerjee I, Agarwal T, Pramanik K, Bhattacharya MK, Pal K. Guar gum and sesame oil based novel bigels for controlled drug delivery. Colloids and Surfaces B: Biointerfaces. 2014;**123**:582-592

Section 2

Nanoemulsions and Spray Drying

Chapter 3

Encapsulation of Natural Bioactive Compounds: Nanoemulsion Formulation to Enhance Essential Oils Activities

Mariem Ben Jemaa, Hanen Falleh and Riadh Ksouri

Abstract

The microencapsulation technology consists of a trap of a compound inside a tiny sphere known as microsphere. The microencapsulation concerns many different active materials such as bioactive compounds, drugs, vitamins, enzymes, flavors, and pesticides. This technology has gained real interest in numerous fields such as agriculture, cosmetic, pharmaceutical, textile, and food. This chapter highlights the encapsulation of essential oils into nanoemulsion-based delivery system as a model for the encapsulation of natural bioactive compounds. Moreover, an investigation of different parameters affecting the stability of produced nanoemulsion was conducted, in addition to the study of the effect of the nanoencapsulation of essential oils on their antibacterial activity. Finally, an enumeration of the advantages of encapsulating essential oils into nanoemulsion-based delivery systems in order to provide a natural food preservatives has been provided.

Keywords: encapsulation, nanoemulsion, essential oil, formulation, stability, antibacterial activity, food preservation

1. Introduction

In recent years, natural antimicrobials attracted consumer attention due to the increased awareness regarding food safety. In this context, new approaches have been adopted in the food preservation field. This includes the use of natural compounds with proven antibacterial activities, like essential oils, as safe preservatives. However, the incorporation of essential oils in foods is not economically and practically ideal. As a matter of fact, essential oils are not only volatile and chemically unstable in the presence of air, light, moisture and high temperatures, but also present hydrophobic properties.

With this respect, the nanoencapsulation of essential oils seems to be an attractive new approach to overcome these impediments.

Since the design of essential-oil-loaded particles is a complex process with interrelated steps [1], choosing encapsulating material and the encapsulation method should be in agreement with the intended matrix in which essential oils are to be introduced [2]. Basically, the nanoencapsulation of essential oils may imply coat, polymeric material, etc. to trap the core material in order to fix listed limitations of using essential oils as natural food preservative. Accordingly, different methods could be adopted for the nanoencapsulation of essential oils, such as nanoemulsion, liposomes, cyclodextrin, etc. In the specific case of essential oil nanoemulsion, the preparation consists on a biphasic liquid system of one liquid solution dispersed in a continuous medium and no polymer shells are used. The immobilization of essential oils in nanoemulsions contributes efficiently to enhance their dispersibility in aqueous solutions, to protect them from interaction with food ingredients, to minimize their impact on the organoleptic properties, as well as to improve their absorption and bioavailability. In this context, numerous researches have been conducted on the nanoencapsulation of essential oils on the purpose of producing a natural powerful food conservator. Gathered data demonstrated that inappropriate formulation, due to a misunderstanding of the process of essential oil encapsulation, can lead to the instability and/or the inefficiency of the produced emulsion.

In this context, the purpose of this chapter is to better understand the phenomenon of encapsulating essential oils into nanoemulsion-based delivery systems. This would widen the knowledge of possible alternatives to consider while designing green food preservatives for future research. Accordingly, this chapter covers firstly a general description of the nanoemulsion delivery systems. Then, an enumeration of common parameters, often used in essential oil nanoemulsion characterization, was conducted. The third part of this chapter involves different adopted methods for the preparation of essential oil nanoemulsion. Moreover, the most relevant parameters affecting the nanoemulsion quality and stability were investigated. Also a special emphasis to the effect of the nanoencapsulation of essential oils on their antibacterial activity was provided. Finally, data on the efficiency of encapsulated essential oils into nanoemulsion-based delivery systems as natural food preservatives have been provided.

2. Nanoemulsion-based delivery system as an example of bioactive compound encapsulation

According to the theory of emulsification, an emulsion is a thermodynamically unstable system consisting of at least two immiscible liquid phases, one of which is dispersed as globules in the other liquid phase [3]. Emulsions can be stabilized by increasing the repulsion between the dispersed and the continuous phases. As a matter of fact, the emulsion formation is a nonspontaneous phenomenon, which requires energy along with the use of emulsifiers. As a matter of fact, emulsifiers are amphiphile molecules that reduce the interfacial tension between the two phases and contribute to the stabilization of dispersed droplets with electrostatic or steric effects [4].

According to the proportion of each used liquid, an emulsion can be considered either as oil in water (O/W) or as water in oil (W/O) emulsion [2]. Indeed, if the oil droplets are dispersed throughout the aqueous phase, the emulsion is called oil-in-water (O/W). In the opposite case, where the water is dispersed as globules in the oil continuous phase, the emulsion is called water-in-oil emulsion (W/O).

It is worthy to mention the increasing interest accorded to multiple emulsions [1]. In this complex type of emulsion system, the W/O or O/W emulsions are dispersed in another immiscible liquid.

Accordingly, O/W/O emulsion is formed by very small oil droplets dispersed in water globules of a W/O emulsion, and W/O/W emulsion is formed by water droplets dispersed in the oil phase of an O/W emulsion (**Figure 1**). Multiple emulsion can be formed by a multistep mechanism. Actually, the deepest drop is formed in the first drop maker and then encapsulated in the next drop maker. In general, multiple emulsions present many advantages such as (1) a good ability to carry both hydrophilic and hydrophobic bioactive ingredients simultaneously, (2) high

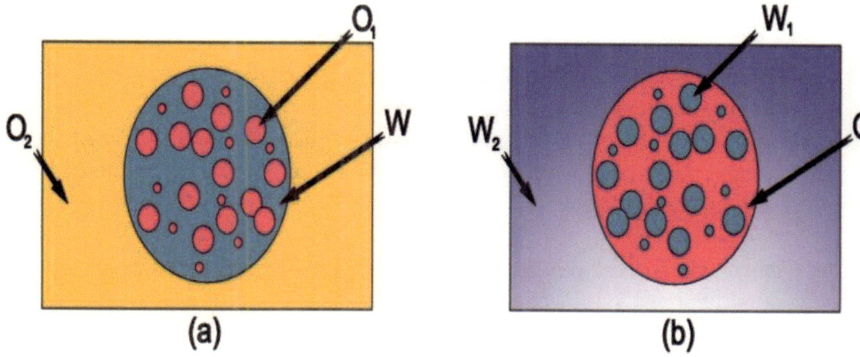

Figure 1.
Schematic representation of the architectures of O/W/O (a) and W/O/W (b) multiple emulsions.

protection of sensitive bioactive molecules from gastrointestinal harsh conditions, and (3) sheltering essential oil's strong taste and smell [1]. Besides their importance, multiple emulsions present some limitations due to their complex structure and their thermodynamic instability [5].

Another special case of emulsion is nanoemulsion. Actually, nanoemulsions are isotropic, clear, and kinetically stable with droplet size inferior to 200 nm [4]. Two types of techniques could be adopted for nanoemulsion preparation: high-energy methods and low-energy methods [6]. This encapsulation method is gaining more and more interest in the scientific community due to its high stability, as compared to emulsions of larger droplet size [7]. Actually, nanoemulsion stability results of its nanoscale droplet size and its large surface area and free energy. With this respect, essential oil nanoemulsion can be formed by the encapsulation of essential oil as the dispersed phase at a nanoscale level.

The main advantages of essential oil nanoemulsions are:

- it possess high kinetic stability;
- solubilize hydrophobic bioactive molecules and enhance their bioavailability;
- can be used for taste masking;
- nontoxic and nonirritant; and
- suitable for human and veterinary use.

In the last few decades, essential oil nanoemulsions have found enormous applications in the field of healthcare, cosmetics, food, agrochemicals, pharmaceuticals, and biotechnology.

2.1 Essential oil nanoemulsion components

The encapsulation of essential oils in nanoemulsion-based delivery system requires basically oil, emulsifier, and aqueous phase:

i. *Oils*: they are used to solubilize the lipophilic bioactive compound and to modulate the viscosity ratio between the dispersed and the continuous phases [8]. The commonly used oils in formulating essential oil food grade nanoemulsions are soyabean oil, ethyl oleate, sesame oil, castor oil, arachis oil, and corn oil.

ii. *Emulsifiers*: emulsifiers are amphiphilic molecules composed of two parts, polar and nonpolar regions [9]. According to their polar group nature, emulsifiers can also be classified into: anionic, cationic, nonionic, and zwitterionic emulsifiers. By lowering the interfacial tension between the two immiscible liquids, emulsifiers contribute significantly in the formulation of essential oil nanoemulsions [9]. Furthermore, they prevent coalescence of newly formed drops.

iii. *Aqueous phase*: the nanoemulsion stability is affected by the nature of the aqueous phase [4]. Particularly, consideration should be given to the aqueous phase pH and presence of electrolytes during nanoemulsion preparation. Theoretically, in essential oil nanoemulsion, continuous phase viscosity could influence droplet size through different mechanisms. It is worthy to mention that the relative importance of these mechanisms depends mainly on the homogenizer design and the used operating conditions [10].

2.2 Nanoemulsion characterization

As detailed in the literature, the majority of the researches, dealing with the essential oil encapsulation into nanoemulsion-based delivery system, have considered that the droplet size and distribution measurements as the most important parameters for their nanoemulsion characterization [4, 6, 11]. Measurements could be determined by laser light scattering, and obtained results are expressed in terms of mean particle size, which is usually represented [11] with Sauter mean diameter, $d_{3,2}$ (expressed in nm), calculated using the following equation:

$$d_{3,2} = (\text{Volume/Surface Area}) = (\Sigma(n_i * d_i)^3)/(\Sigma(n_i * d_i)^2) \qquad (1)$$

where n_i is the number of droplets and d_i is the droplet diameter.

In addition to these two listed parameters, the stability of essential oil nanoemulsion, which means its ability to resist physicotemporal changes, could be investigated by storing nanoemulsions at different temperatures and periods and measuring at each time point the droplet size variation [12]. Also, some researches have measured the viscosity, the density, the color, the turbidity as physical characterization of their nanoemulsion [13, 14], while others have focused their researches on the investigation of the biological activities of produced nanoemulsions [11, 15].

2.3 Methods of nanoemulsion preparation

The laborious step of formulation aims to produce stable nanoemulsion. Indeed, any emulsion system, if inappropriately formulated, may be subject to a variety of physicochemical phenomena, which can seriously affect the stability and the biological efficiency of the produced nanoemulsion [16].

A well-homogenized and stable emulsion whose droplet diameters figure in the nanoscale level can only be obtained under a complex alliance between physical and chemical forces [17]. As a matter of fact, the exclusive use of physical forces remains often insufficient to obtain stable nanoemulsions. The aim role of physical forces is to reduce the size of the dispersed phase droplets at a certain level, while chemical interactions between different medium components interfere to maintain newly formed droplets from fusing together.

2.3.1 Chemical forces

From a chemical point of view, the interfacial tension remains a critical parameter in the process of essential oil nanoemulsification. As a matter of fact, interfacial tension is known as the inward attraction of molecules at the surface of immiscible liquids due to the imbalance of their attractive forces [18]. With this respect, the nanoemulsion formation can occur only if the interfacial tension between the two immiscible phases decreased sufficiently to assure their mixture [11]. To reach such change in interfacial tension, an appropriate amount of an appropriate emulsifier should be used to surround and stabilize all neo-formed nanodroplets. Indeed, an emulsifier is not only able to reduce the interfacial tension of the two immiscible phases, but also it presents an effective stabilizer for the newly formed droplets [19]. In this way, the surfactants convert large globules into small ones and avoid small globules from coalescing into large ones, by reducing the repellent force between the liquids and withdrawing the attraction of liquids for their own molecules [20]. It is worthy to mention that to ensure its fundamental role in the nanoemulsification process, surfactants should be used at a higher concentration than its critical micellar concentration "CMC." As a matter of fact, the increase of surfactant concentrations on oil-water interface increases the adsorption of surfactant molecules, leading to the improvement of their ability to reduce the interfacial tension. Once the adsorption saturation is reached at the oil-water interface, the adsorption of surfactant molecules would stop increasing; thus, the interfacial tension remains constant [21].

2.3.2 Physical forces

Physical emulsification is one of the most crucial steps in nanoencapsulation process since it affects deeply the quality of the final emulsion (encapsulation efficiency, nanoemulsion stability, or biological efficacy). Different homogenization methods, as detailed in **Table 1**, can be used such as high-pressure homogenization, ultrasonic homogenization, and microfluidization [22].

Devices	Encapsulated bioactive compound	Obtained particle size (nm)	Disadvantages	References
High-pressure homogenization	*Thymus capitatus* essential oil	110	- High production costs	[11]
	Melaleuca alternifolia essential oil	175		[2]
Microfluidization	*Eugenia caryophyllata* essential oil	21	- High production costs - Less aseptic processing	[23]
	Lemon myrtle essential oil	97		[15]
Sonication	Rosemary essential oil	187	- Heat generation during the process due to high shear forces and cavitations	[24]
	Thymus vulgaris essential oil	121		[25]

Table 1.
The different devices often used for the nanoencapsulation of essential oils.

High-pressure homogenization: the use of high-pressure (30–350 MPa) homogenization can reduce an emulsion droplet size to a nanoscale level and improve its stability by the reduction of the creaming rate. In this method, two immiscible liquids along with the used emulsifier are forced to pass through a small orifice where their mixture is subjected to a turbulence and shear flow intense levels [2]. All of which leads to the break-up of the dispersed phase into small droplets.

Microfluidization: this technic involves the transfer of mechanical energy to fluid particles under a high-pressure environment. More specifically, immiscible liquids are pumped and split into two opposite microstreams, which are impacted or collided against each other in a chamber, called the interaction chamber, where shear, turbulent and cavitation forces are generated using high-pressure displacement pump [9].

Ultrasonication: the droplet formation at a nanoscale level using ultrasound homogenization mechanisms is mainly based on cavitation, where ultrasound waves hit the liquid surface and form high-velocity jets. To do that, a probe sonicator is brought in contact with the dispersion of emulsifier and liquids. The generated mechanical vibration and cavitation can provide the necessary energy input for the formation of small sized droplets [26].

2.4 Parameters affecting nanoemulsion droplet size and stability

The droplet homogeneity, and consequently emulsion stability, depends basically on the physical characteristics of each used component for the formation of essential oil nanoemulsion [9]. **Table 2** summarizes the obtained droplet size of different tested formulation:

2.4.1 Dispersed phase characteristics

During the process of encapsulating essential oils into nanoemulsion-based delivery system, dispersed phase characteristics influence profoundly the final product properties [3]. In this context, the physical characteristics of the encapsulated essential oil (such as interfacial tension and viscosity) present a key factor in the nanoemulsion stability [11]. For example, it is difficult to nanoencapsulate pure fixed oils due to their high viscosity. Indeed, if the dispersed phase viscosity increases, it will become more difficult to breakup within the high-pressure homogenizer. As a result, nanoemulsions with larger droplets will be formed. This phenomenon was also confirmed by other researches for various types of homogenization device, who declared that the droplet breakup becomes easier as the viscosity of the dispersed phase decreases [30]. Also, the increase of the dispersed phase viscosity generated a significant increase of the mean droplet diameter, from around 92 to around 125 nm [9]. Other researches demonstrated that instantaneous unstable emulsions were obtained while trying the nanoencapsulation of pure essential oil [11]. These

Encapsulated essential oil	Dispersed phase	Emulsifier	Droplet size (nm)	References
Thymus capitatus	Mixture with 30% of soybean oil	SDS	110	[11]
Eucalyptus globulus	Pure essential oil	Tween 20	60	[27]
Carvacrol	Mixture with 60% of medium chain triglyceride	Tween 80	100	[28]
Peppermint	Pure essential oil	Tween 80	70	[29]

Table 2.
Formulation effect on the droplet size of essential oil nanoemulsion.

findings were explained by the lower viscosity of pure essential oils. Actually, to obtain a stable emulsion with a nanoscale droplet size, there is an optimum range of disperse-to-continuous phase viscosity ratios [10]. With this respect, it would be convenient to nanoencapsulate a mixture of essential oil and fixed oil as dispersed phase in order to obtain an essential oil nanoemulsion.

2.4.2 Continuous phase characteristics

Changing the continuous phase viscosity influences the nanoemulsion droplet size [9]. More specifically, the increase of the continuous phase viscosity leads to the droplet diameter decrease [10]. Actually, the increase of the continuous phase viscosity induces the increase of the disruptive shear stresses, which leads to the increase of droplet fragmentation.

2.4.3 Emulsifier chemical characteristics

Although a wide range of molecules may be used in the essential oil nanoemulsification (exp. colloids, special particles), only the emulsifier case will be presented in this chapter. Different systems were adopted in order to classify surfactants. For instance, a very useful system for the classification of surfactants is standardized on the basis of their solubility in water. In this system, numerical values are called the hydrophilic-lipophilic balance (HLB), which involves the relative affinity of the surfactant for water and oil. The HLB is defined as the relative efficiency of the hydrophilic portion of the surfactant molecule to its lipophilic portion. It is worthy to mention that emulsifiers with HLB values ranged between 3 and 6 are usually used for w/o emulsions. Whereas emulsifiers with HLB values ranged between 7 and 20 are used for o/w emulsions [5]. Besides, other researchers have based their investigations on the effect of emulsifier chemical nature in producing homogenous nanoemulsions [9, 11, 31]. Their findings confirmed that emulsifier type presents significant impact on the final emulsion stability. This variation of emulsifier behaviors could be explained by the difference in the stabilization process of each one. Previous researches demonstrated that under similar homogenization conditions, small-molecule emulsifiers (exp. Tween and SDS) can be more effective to make small droplets than biopolymers (exp. caseinate and β-lactoglobulin) due to their rapid adsorption to the droplet surfaces [9]. Moreover, charged emulsifiers can be more efficient in producing homogenous nanoemulsions as compared to nonionic ones [11]. In fact, contrary to nonionic emulsifier, which uses steric repulsion to stabilize the dispersed phase, charged emulsifiers use their electrostatic repulsion. Actually, for nonionic emulsifier-based emulsion, tails envelop essential oil inside the droplet [32, 33]. In this case, the hydrophilic nature of tails may repel the hydrophobic essential oils leading to a significant heterogeneity of the droplet diameters. In anionic emulsifier-based emulsions, the inverse occurs. As a matter of fact, the adsorption of negatively charged heads of emulsifier molecules to oil droplets surface increases the electrostatic repulsion between droplets, leading to the formation of a stable nanoemulsion [34]. In this case, the charged heads of SDS molecules envelop essential oil inside the droplet, while the hydrophilic tails stay outside leading to an appropriate homogeneous nanoemulsion.

2.5 Nanoemulsion instability

Nanoemulsions lose their stability, which is an irreversible phenomenon in nature, in a very large time frame that may vary from few minutes to several years, depending on formulation and storage conditions [35]. In this context, nanoemulsions stability can be checked according to their droplet size growth and their appearance.

Generally, emulsion stability depends mainly on emulsifier behavior, emulsion composition, and its droplet size distribution [3]. Indeed, nanoemulsions, due to their characteristic nanoscale droplets size, exhibit higher stability against creaming or sedimentation, than emulsions with larger droplet diameters. Actually, diffusion rate and Brownian motion exhibited by nanoemulsion droplets predominates over the sedimentation or the creaming rate [4].

Concerning the emulsifier behavior, nanoemulsions prepared using nonionic surfactants do not usually flocculate, as no attractive forces are created [11].

Also, nanoemulsion stability depends strongly on their storage time and conditions. Actually, small droplets of freshly made nanoemulsion could initially be distributed in the medium, but are rather unstable, resulting in droplet growth during long storage, and these new large droplets are the source of flocks. In fact, droplets flocculation appears whenever the interfacial tension of the dispersed phase is weaker than its own net attractive forces [36]. Accordingly, nanoemulsions storage temperature increase not only provokes the increase in molecules thermal agitation [37], but also the decrease in their interfacial tension. Consequently, the droplet diameter of instable thermodynamic nanoemulsions would tend to increase to reduce medium total free energy. The nanoemulsion storage at high temperature (up to 55°C) generated new populations of larger droplets after 15 storage days [9].

Worthy to note that nanoemulsion instability occurs due to alteration in droplet size through mechanisms such as Coalescence and Ostwald ripening.

i. *Coalescence:* this phenomenon results from the fusion of two or more droplets into one larger one by the breakdown of the thin film existing between them [38]. As a matter of fact, coalescence occurs if the adhesion force between two droplets exceeds the turbulent force that creates the dispersion. Coalescence can be prevented by the addition of emulsifiers, which have the same charges causing repulsion between two droplets [4].

ii. *Ostwald ripening:* this phenomenon is characterized by change in droplet size and distribution, as well as by turbidity apparition in nanoemulsions [3]. Actually, Ostwald ripening occurs with time passage due to the migration of droplets from the dispersed phase (high Laplace pressure) to the continuous phase (low Laplace pressure), leading to molecular diffusion [39]. To prevent Ostwald ripening, several parameters should be taken in consideration such as: the physical properties of the bioactive compound, the mutual solubility of the phases, the nature and concentration of used emulsifier, preparation methods, and storage conditions [4].

Otherwise, practically nanoemulsions are usually stored at lower temperatures, inducing therefor, their longer stability and higher resistance to droplet aggregation.

2.6 Effect of the nanoencapsulation process on essential oils antibacterial efficiencies

As the antibacterial efficiency is a fundamental characteristic of essential oils, different methods were adopted in order to seek the effect of the nanoencapsulation process on essential oil antibacterial potency [11, 25, 40]. All results depicted clear amelioration in the antibacterial efficiency. Such amelioration suggests a refinement of the mode of action of essential oils after their nanoencapsulation in fighting pathogenic bacteria. Actually, some researchers considered the nanoemulsion as a transporter for essential oils to cross the bacterial cellular membrane, allowing them to overcome their hydrophobic limitation [11]. Indeed, essential oils exert their known antibacterial effect from the inner side of the cytoplasmic membrane [28]. Essential oil antibacterial effect is based on their abilities to disrupt the bacterial cytoplasmic membrane to lose its properties as a barrier, matrix for enzymes, and energy transducer, all of which will compromise the cell viability leading to its death [41, 42]. However, essential oil presents low water solubility, inducing its rough distribution in the medium, which can limit its antibacterial action [28, 42].

Moreover, it is worthy to note that the interesting antibacterial activity of nanoencapsulated essential oil could present a promising procedure to fight against the global issue involving drug-resistant strains. As a matter of fact, in addition to the antibacterial activity amelioration of essential oil after their nanoencapsulation, many essential oils have succeeded to surpass the efficiency of current antibiotic. These findings are very promoting, especially that the development of drug-resistant strains has become a worldwide concern [43]. As a matter of fact, typical antibiotic killing technique consists on blocking bacterial ribosome formation at the initiation step [44]. Consequently, no protein synthesis, which is obligatory for bacterial metabolism and survival, takes place leading the bacterial cell death. In this context, antibiotic-resistant strain could be formed after mutation of the initial bacteria whose ribosome formation continues even in the presence of the drug. However, the efficiency of nanoencapsulated essential oil, based on the nonspecific disruption of bacterial cell membranes, can resist bacterial mutation and maintain its bactericidal activities. Accordingly, the use of nanoencapsulated essential oils into nanoemulsion-based delivery system is very efficient to fight pathogenic bacteria and would not conduct the development of resistant strains, which could remediate to the bacterial resistance problem, caused by the widespread and inappropriate use of antibiotics [11].

2.7 Nanoencapsulated essential oils as efficient food conservators

It has been repeatedly demonstrated that essential oil efficiency can be reduced in real food systems due to their hydrophobic character and their low solubility in water as compared to *in vitro* model system [11, 45]. This reduce in essential oil activity is more noticed in foods with high fat level such as milk, mayonnaise, butter, etc. [46]. For instance, it has been demonstrated that in cheese, an increase up to 100-fold of the essential oil concentration was required to assure comparable antimicrobial efficacy of the *in vitro* model system [47]. The difference in the essential oil antimicrobial efficiency between *in vitro* and real food system can be attributed to the essential oil dissolution in the lipid phase of an ailment, inducing the decrease of their concentration in the aqueous phase [48]. However, it is in the aqueous phase where pathogenic bacteria typically proliferate [46]. Consequently, to ensure similar antimicrobial activity in *in vitro* and in real food system, essential oils have to be relocated in the aqueous phase of the food, in order to be in continuous contact with the pathogenic microorganisms. In order to accomplish this goal, the encapsulation of essential oils into a nanoemulsion-based delivery system seems to be an interesting approach. As a matter of fact, the hydrophilic outer surface of the nanoemulsion enables essential oils to stay in the food's aqueous phase, while its hydrophobic inner core ensures its harbor.

Also, the nanoencapsulation of essential oils prevents their interaction with food components, which induce a positive impact on essential oil antimicrobial efficiency. Indeed, bulk essential oils tend to bind with hydrophobic food molecules leading to the reduction of their availability to fight pathogenic microorganisms [48]. In contrast, nanodispersed essential oil is evenly distributed in food matrix and can be released locally to keep its concentration sufficiently high to inhibit the growth of the spoilage bacteria [49].

Moreover, designing systems that entrap essential oil molecules can reduce the adverse interaction of their characteristic aroma with the original food flavor. As a matter of fact, the use of essential oils as food conservatives can be limited by their sensorial impact on the final food product. Accordingly, adding *Lavandula* and *Chamaemelum* spp. essential oils to yoghurt decreased its acceptability by panelist [50]. Actually, when incorporated into food system, bulk essential oils bind with fats [51]. Such bindings could alter the sensory appreciation of the incorporated aliment, since the taste appreciation depends mainly on its fat quality [52]. On the other way, the encapsulation of *Thymus capitatus* essential oil, into a nanoemulsion-based delivery system, ameliorated significantly its sensorial impact when incorporated into milk [48]. Authors explained their findings by the fact that, when nanoencapsulated, essential oil components were trapped inside droplets and were not able to interact with milk ingredients [53]; therefore, their incorporation would not modify fat quality.

3. Conclusion

The encapsulation of natural bioactive compounds into a nanoemulsion-based delivery system presents definitely an interesting approach to facilitate and ameliorate the valorization of essential oils as natural and green food conservators. Actually, the nanoencapsulation of essential oils protect them from brutal external conditions, ameliorate their distribution in the medium leading to an amelioration of their bactericidal potency, as well as prevent their interaction with food components, which induce a positive impact on their incorporation efficacy. However, special attention should be attributed to the formulation step of essential oil nanoemulsion to avoid different physicochemical phenomena, which can seriously affect the stability and the biological efficiency of the produced nanoemulsion.

Acknowledgements

This study was supported by the Tunisian Ministry of Higher Education, Scientific Research, Information, and Communication Technologies.

Conflict of interest statement

The authors declare that there are no conflict of interests.

Funding sources

This research did not receive any specific grant from funding agencies in the public, commercial, or not-for-profit sectors.

Author details

Mariem Ben Jemaa*, Hanen Falleh and Riadh Ksouri
Laboratory of Aromatic and Medicinal Plants, Biotechnology Center of Borj Cedria, Hammam-Lif, Tunisia

*Address all correspondence to: mariembenjemaa@yahoo.fr

References

[1] Sagiri SS, Anis A, Pal K. A review on encapsulation of vegetable oils: Strategies, preparation methods and applications. Polymer-Plastics Technology and Engineering. 2015;**55**(3):291-311. DOI: 10.1080/03602559.2015.1050521

[2] Donsi F, Annunziata M, Sessa M, Ferrari G. Nanoencapsulation of essential oils to enhance their antimicrobial activity in foods. LWT—Food Science Technology. 2011;**44**(9):1908-1914. DOI: 10.1016/j.lwt.2011.03.003

[3] Tadros T, Izquierdo P, Esquena J, Solans C. Formation and stability of nanoemulsions. Advance in Colloid and Interface Science. 2004;**108**:303-318. DOI: 10.1016/j.cis.2003.10.023

[4] Setya S, Talegaonkar S, Razdan BK. Nanoemulsions: Formulation methods and stability aspects. World Journal of Pharmacy And Pharmaceutical Sciences. 2014;**3**(2):2214-2228

[5] Khan BA, Akhtar N, Khan HMS, Waseem K, Mahmood T, Rasul A, et al. Basics of pharmaceutical emulsions: A review. African Journal of Pharmacy and Pharmacology. 2011;**5**(25): 2715-2725. DOI: 10.5897/AJPP11.698

[6] Gupta A, Burak Eral H, Alan Hattona T, Doyle PS. Nanoemulsions: Formation, properties and applications. The Royal Society of Chemistry. 2016;**12**:2826-2841 DOI: 10.1039/c5sm02958a

[7] Qadir A, Faiyazuddin MD, Talib Hussain MD, Alshammari TM, Shakeel F. Critical steps and energetics involved in a successful development of a stable nanoemulsion. Journal of Molecular Liquids. 2016;**214**:7-18. DOI: 10.1016/j.molliq.2015.11.050

[8] Zhang Y, Gao J, Zheng H, Zhang R, Han Y. The preparation of 3,5-dihydroxy-4-isopropylstilbene nanoemulsion and in vitro release. International Journal of Nanomedecine. 2011;**6**:649-657. DOI: 10.2147/IJN.S17242

[9] Qian C, McClements DJ. Formation of nanoemulsions stabilized by model food-grade emulsifiers using high-pressure homogenization: Factors affecting particle size. Food Hydrocolloids. 2011;**25**:1000-1008. DOI: 10.1016/j.foodhyd.2010.09.017

[10] Håkansson A, Tragårdh C, Bergenståhl B. Studying the effects of adsorption, recoalescence and fragmentation in a high-pressure homogenizer using a dynamic simulation model. Food Hydrocolloids. 2009;**23**(4):1177-1183. DOI: 10.1016/j.foodhyd.2008.10.003

[11] Ben Jemaa M, Neves MA, Falleh H, Isoda H, Ksouri R, Nakajima M. Nanoencapsulation of *Thymus capitatus* essential oil: Formulation process, physical stability characterization and antibacterial efficiency monitoring. Industrial Crops and Products. 2018;**113**:414-421. DOI: 10.1016/j.indcrop.2018.01.062

[12] Rodríguez-Rojo S, Visentin A, Maestri D, Cocero MJ. Assisted extraction of rosemary antioxidants with green solvents. Journal of Food Engineering. 2012;**109**(1):98-103. DOI: 10.1016/j.jfoodeng.2011.09.029

[13] Basak S, Guha P. Betel leaf (*Piper betle* L.) essential oil microemulsion: Characterization and antifungal activity on growth, and apparent lag time of *Aspergillus flavus* in tomato paste. LWT–Food Science and Technology. 2017;**75**:616-623. DOI: 10.1016/j.lwt.2016.10.021

[14] Ghosh V, Mukherjee A, Chandrasekaran N. Formulation and

characterization of plant essential oil based nanoemulsion:Evaluation of its larvicidal activity against *Aedes aegypti*. Asian Journal of Chemistry. 2013;**25**:321-323

[15] Buranasuksombat U, Kwon YJ, Turner M, Bhandari B. Influence of Emulsion Droplet Size on Antimicrobial Properties. Food Science and Biotechnology. 2011;**20**(3):793-800. DOI: 10.1007/s10068-011-0110-x

[16] Chang Y, McLandsborough L, McClements DJ. Physical properties and antimicrobial efficacy of thyme oil nanoemulsions: Influence of ripening inhibitors. Journal of Agricultural and Food Chemistry. 2012;**60**:12056-12063. DOI: 10.1021/jf304045a

[17] Wooster TJ, Golding M, Sanguansri P. Impact of oil type on nanoemulsion formation and Ostwald ripening stability. Langmuir. 2008;**24**(22): 12758-12765. DOI: 10.1021/la801685v

[18] John EH. Dispensing of Medication. Pennysylvania, USA: Mack Publishing Company; 1976. pp. 345-358

[19] Walstra P. Principles of emulsion formation. Chemical Engineering Science. 1993;**48**:333-349. DOI: 10.1016/0009-2509(93)80021-H

[20] Paul BK, Mitra RK. Water solubilization capacity of mixed reverse micelles: Effect of surfactant component, the nature of the oil, and electrolyte concentration. Journal of Colloids and Interface Science. 2005;**288**:261-279. DOI: 10.1016/j.jcis.2005.02.088

[21] Yuan C, Pu WF, Wang XC, Sun L, Zhang Y, Cheng S. The effects of interfacial tension, emulsification and surfactant concentration on oil recovery in surfactant flooding process for high temperature and high salinity reservoirs. Energy and Fuels. 2016;**55**(48):12387-12397. DOI: 10.1021/acs.energyfuels.5b01393

[22] McClements DJ. Nanoemulsions versus microemulsions: Terminology, differences, and similarities. Soft Matter. 2012;**8**:1719-1729. DOI: 10.1039/C2SM06903B

[23] Salvia-Trujillo L, Rojas-Graü A, Soliva-Fortuny R, Martín-Belloso O. Physicochemical characterization and antimicrobial activity of foodgrade emulsions and nanoemulsions incorporating essential oils. Food Hydrocolloids.2014;**43**:547-556. DOI: 10.1016/j.foodhyd.2014.07.012

[24] Koh LL, Chandrapala J, Zisu B, Martin GJ, Kentish SE, Ashokkumar M. A comparison of the effectiveness of sonication, high shear mixing and homogenisation on improving the heat stability of whey protein solutions. Food Bioprocess Tehnology. 2014;**7**:556-566. DOI: 10.1007/s11947-013-1072-1

[25] Ghaderi-Ghahfarokhi M, Barzegar M, Sahari MA, Azizi MH. Nanoencapsulation approach to improve antimicrobial and antioxidant activity of thyme essential oil in beef burgers during refrigerated storage. Food Bioprocess Technology. 2016;**9**(7):1187-1201. DOI: 10.1007/s11947-016-1708-z

[26] Delmas T, Piraux H, Couffin AC, Texier I, Vinet F, Poulin P, et al. How to prepare and stabilize very small nanoemulsions. Langmuir. 2010;**27**(5):1683-1692. DOI: 10.1021/la104221q

[27] Sugumar S, Mukherjee A, Chandrasekaran N. Nanoemulsion formation and characterization by spontaneous emulsification: Investigation of its antibacterial effects on *Listeria monocytogenes*. Asian Journal of Pharmaceutics. 2015;**9**:23-28. DOI: 10.22377/ajp.v9i1.427

[28] Landry KS, Chang Y, McClements DJ, McLandsborough L. Effectiveness of a novel spontaneous carvacrol nanoemulsion against *Salmonella enterica Enteritidis* and *Escherichia coli* O157:H7 on contaminated mung bean and alfalfa seeds. International Journal of Food Microbiology. 2014;**187**:15-21. DOI: 10.1016/j.ijfoodmicro.2014.06.030

[29] Barzegar H, Mehrnia MA, Nasehi B, Alipour M. Fabrication of peppermint essential oil nanoemulsions by spontaneous method: Effect of preparing conditions on droplet size. Flavor and fragrance Journal. 2018;**33**(5):351-356. DOI: 10.1002/ffj.3455

[30] Jafari SM, Assadpoor E, Bhandari B, He Y. Nano particle encapsulation of fish oil by spray drying. Food Research International. 2008;**41**(2):172-183. DOI: 10.1016/j.foodres.2007.11.002

[31] Reynhout XEE, Hoekstra L, Meuldijk J, Drinkenburg AAH. Contribution of steric and electrostatic repulsion forces to the stability of styrene latices copolymerized with acrylic acid. Steric and Electrostatic Repulsion Forces. 2003;**41**(19):2985-2995. DOI: 10.1002/pola.10889

[32] Nash JJ, Erk KA. Stability and interfacial viscoelasticity of oil-water nanoemulsions stabilized by soy lecithin and tween 20 for the encapsulation of bioactive carvacrol. Colloids and Surfaces A: Physicochemistry and Engenieering Aspects. 2017;**517**:1-11. DOI: 10.1016/j.colsurfa.2016.12.056

[33] Li X, Li H, Xiao Q, Wang L, Wang M, Lu X, et al. Two-way effects of surfactants on Pickering emulsions stabilized by the self-assembled microcrystals of α-cyclodextrin and oil. Physical Chemistry Chemical Physics. 2014;**16**:14059-14069. DOI: 10.1039/c4cp00807c

[34] Krstonosic V, Dokic L, Nikolic I, Dapcevc T, Hadnadev M. Influence of sodium dodecyl sulphate (SDS) concentrations on the disperse and rheological characteristic of oil in water emulsion stabilized by octenyl succinic anhydride modified starch-SDS mixture. Journal of Sebian Chemistry Society. 2012;**77**:83-94. DOI: 10.2298/JSC100630150K

[35] Reza A. Nanometric-scaled emulsions (nanoemulsions). Iranian Journal of Pharmaceutic Research. 2010;**9**(4):325-326

[36] Kim HJ, Decker EA, McClements DJ. Impact of protein surface denaturation on droplet flocculation in hexadecane oil-in-water emulsions stabilized by beta-lactoglobulin. Journal of Agricultural and Food Chemistry. 2002;**50**(24):7131-7137. DOI: 10.1021/jf020366q

[37] Taylor L. The Healing Power of Rainforest Herbs. A Guide to Understanding and Using Herbal Medicinals. New York: Square One Publishers; 2005

[38] Delmas T, Piraux H, Couffin AC, Texier I, Vinet F, Poulin P, et al. How to prepare and stabilize very small nanoemulsions. Langmuir. 2010;**27**(5):1683-1692

[39] Tadros TF. Emulsion science and technology: A general introduction. In: Tadros TF, editor. Emulsion Science and Technology. Weinheim: Wiley-VCH; 2009. DOI: 10.1002/9783527626564.ch1

[40] Bilia AR, Santomauro F, Sacco C, Bergonzi MC, Donato R. Essential oil of *Artemisia annua* L.: An extraordinary component with numerous antimicrobial properties. Trends Plant Science. 2014;**9**:97-104. DOI: 10.1155/2014/159819

[41] Turina AV, Nolan MV, Zygadlo JA, Perillo MA. Natural terpenes: Self-assembly and membrane partitioning.

Biophysical Chemistry. 2006;**122**: 101-113. DOI: 10.1016/j.bpc.2006.02.007

[42] Gill AO, Holley RA. Disruption of *Escherichia coli*, *Listeria monocytogenes* and *Lactobacillus sakei* cellular membranes by plant oil aromatics. International Journal of Food Microbiology. 2006;**108**:1-9. DOI: 10.1016/j.ijfoodmicro.2005.10.009

[43] Zaman SB, Hussain MA, Nye R, Mehta V, Mamun KT, Hossain N. A review on antibiotic resistance: Alarm bells are ringing. Cureus. 2017;**9**(6):1403-1411. DOI: 10.7759/cureus.1403

[44] Luzzatto L, Apirion TD. Schlessinger D, Mechanism of action of Streptomycin in *E. coli*: Interruption of the ribosome cycle at the initiation of protein synthesis. Proceedings of the National Academy of Sciences. 1968;**60**:873-880

[45] Bonnaud M, Weiss J, McClements DJ. Interaction of a food-grade cationic surfactant (lauric arginate) with food-grade biopolymers (pectin, carrageenan, xanthan, alginate, dextran, and chitosan). Journal of Agricultural and Food Chemistry. 2010;**58**:9770-9777. DOI: 10.1021/jf101309h

[46] Edris AE. Pharmaceutical and therapeutic potentials of essential oils and their individual volatile constituents: A review. Phytotherapy Research. 2007;**21**:308-323. DOI: 10.1002/ptr.2072

[47] Mendoza-Yepes MJ, Sanchez-Hidalgo LE, Maertens G, Marinlniesta F. Inhibition of *Listeria monocytogenes* and other bacteria by a plant essential oil (DMC) in Spanish soft cheese. Journal of Food Safety. 1997;**17**:47-55. DOI: 10.1111/j.1745-4565.1997.tb00175.x

[48] Ben Jemaa M, Falleh H, Neves MA, Isoda H, Nakajima M, Ksouri R. Quality preservation of deliberately contaminated milk using thyme free and nanoemulsified essential oils. Food Chemistry. 2017:726-734. DOI: 10.1016/j.foodchem.2016.09.030

[49] Shah B, Davidson PM, Zhong Q. Nanodispersed eugenol has improved antimicrobial activity against *Escherichia coli* O157:H7 and *Listeria monocytogenes* in bovine milk. International Journal of Food Microbiology. 2013, 2013;**161**: 53-59. DOI: 10.1016/j.ijfoodmicro

[50] Ghalem BR, Zouaoui B. Evaluation of the quality of steamed yogurt treated by *Lavandula* and *Chamaemelum* species essential oils. Journal of Medicinal Plants Research. 2013;**7**(42):3121-3126. DOI: 10.5897/JMPR12.1037

[51] Shah GM, Groot JCJ, Oenema O, Lantinga EA. Covered storage reduce losses and improve crop utilisation of nitrogen from solid cattle manure. Nutrient Cycling in Agroecosystems. 2012;**94**:299-312. DOI: 10.1007/s10705-012-9543-8

[52] Guinard JX, Mazzucchelli R. Effects of sugar and fat on the sensory properties of milk chocolate: Descriptive analysis and instrumental measurements. Journal of the Science of Food and Agriculture. 1999;**79**: 1331-1339. DOI: 10.1002/(SICI)1097-0010(199908)79:11<1331::AID-JSFA365>3.0.CO;2-4

[53] Selim S. Antimicrobial activity of essential oils against vancomycin-resistant *Enterococci* (vre) and *Escherichia coli* o157:h7 in feta soft cheese and minced beef meat. Brazilian Journal of Microbiology. 2011;**42**:187-196. DOI: 10.1590/S1517-83822011000100023

Chapter 4

Microencapsulation Techniques of Herbal Compounds for Raw Materials in Food Industry, Cosmetics and Pharmaceuticals

Tri Yuni Hendrawati, Alvika Meta Sari, Muhamad Iqbal Syauqi Rahman, Ratri Ariatmi Nugrahani and Agung Siswahyu

Abstract

Microencapsulation is a technique or process of wrapping very small gas particles, gases, or active solid content with a coating material/membrane to protect the active particles (core) from environmental influences like unwanted effects such as light, moisture, and oxygen to increase shelf life of the product. Microencapsulation proposes to protect sensitive food components, reduce nutritional losses, expand the usefulness of sensitive food components, add certain food to other food, protect flavors and fragrances, convert liquid food components to more convenient solids handled, and protect materials from environmental influences. Product microcapsulation can be used as raw material for the food industry, cosmetics, and pharmaceuticals using bioactive compounds. From the results of the curcuminoid content testings, it can be observed that an increase of drying temperature produces lower amount of curcuminoid contents, which is caused by the inability of curcuminoid compounds to be preserved by maltodextrin, as the microencapsulant. The best temperature to preserve curcuminoid compounds is at 110°C, in which 10.52% is preserved. Hence, for *Aloe vera* processing, the optimum drying temperature was 120°C which maintained the active component of *Aloe vera* powder such as *Aloenin (B)*, *Aloeresin A*, and *Chrysophanol*.

Keywords: microencapsulation, herbal compounds, maltodextrin, *Aloe vera*, cosmetics

1. Introduction

Microencapsulation is an encapsulation technique or process of very small gas particle, gas, or active solid substance with coating/membrane materials with the purpose of protecting the active particle (core) from unwanted environmental influences, such as radiation, humidity, and oxidation to increase shelf life [1]. These capsules are measured in one (1) micron (1/1000 mm) to seven (7) mm, and release their contents at a measured time according to their applications [2].

Microencapsulation aims to protect sensitive food particle, reduce loss of nutrition, expand the uses of sensitive food material, add certain food particles into other food materials, protect tastes and aroma, modify the state of food material from liquid to solid for ease of handling, and protect food particles from environmental effects. Protection provided by microencapsulation can also prevent degradation caused by radiation or oxidation, and also slow down evaporation on volatile compounds [3].

The results of a microencapsulation process are microcapsules containing an active compounds or raw materials surrounded by membrane or cell. The material encapsulated is usually referred to as the core, internal phase, or insert. The coating material is called coat, encapsulant, or shell with varied number and thickness. Coat, shell, encapsulant, or wall is designed to protect the core from destructive factors such as radiation, oxygen, and humidity. In microencapsulation, capsule is designed and prepared to achieve all the needs considering the natures of the core or coating materials, the desired usage of the material, and storage condition [2].

Encapsulants from carbohydrates, such as maltodextrin, starch, and arabic/acacia gum are widely used. However, these materials generally have weak surface tension and require modification or are used with agents with active surface tension to encapsulate oil-based substances [4].

There are four mechanisms of core release from microcapsules: degradation, dissolution, and melting of capsule walls, and diffusion of core materials through broken shell. Abrasion (slow erosion of capsule shell) and biodegradation are two other mechanisms that are less frequently employed [5].

The use of microencapsulation technology has been applied in many fields, such as drug encapsulation in the pharmaceutical industry, adhesive materials, agrochemicals, live cells, catalysts, vitamin storage, and so on. The advantages of microencapsulation are handling liquid as solid, preserving aroma or taste effectively in the food industry, protecting core substances from detrimental effects of the environment, safe handling of toxic materials, and controlling the delivery of drugs [2].

The benefits of microencapsulations are preserving the functions of active compounds, extending shelf life, covering unpleasant taste or aroma (unpleasant taste but high benefits), facilitating handling, facilitating control, improving appearance, and improving taste and colors. Microencapsulation can be prepared by emulsified coating or fluidized bed coating. Microencapsulation process with spray dryer method consists of two phases: oil emulsification in polymer solution and solvent removal using hot air. The polymers used are from many kinds of polysaccharides and proteins, such as starch, arabic gum, gelatin, albumin, and casein [4].

In an emulsification, emulsion is formed when minute oil droplets are dispersed in an emulsifier, in this case a polymer. Emulsion is a mixture system containing two immiscible liquid phases, in which one phase is dispersed in the other phase in the form of droplets. Almost in all food products, the diameters of the droplets range from 0.1 to 100μm. Emulsion is an unstable system in which the phases tend to separate. In an emulsion system consisting of pure oil and pure water, it is easy to for two layers based on the difference in densities. This phenomenon is caused by the tendency of the droplets to combine with nearby droplets and often produce a perfect separation. As such, stability is one of important factors in the encapsulation process using spray dryer. The process to make two immiscible solutions form an emulsion is called homogenization and the mechanism to perform this process is called homogenizer. To differentiate between the natural state from the initial components, homogenization can be more appropriately categorized as primary (emulsion formation) and secondary homogenizations (droplet size reduction) [4].

In almost all microcapsules, the coating materials are usually made of organic polymers, although wax and fats have been used, especially in the uses for food and pharmaceutical products, the coating materials have to meet the specifications required by the FDA [4].

Microencapsulation process can be performed with several techniques, such as spray drying, spray cooling, extrusion, and coacervation [3]. Out of those four methods, spray drying is most frequently employed. Spray drying has become the most important method in the water removal process (dehydration) for liquid food products in the western world. This dehydrator is a diabetic dehydrator, and there are many considerations on solid-state diabetic dehydrator that can be applied. This process is a conversion from a liquid state into dry particles by spraying materials into the hot dehydrating medium. The dry products from this dehydrating process can be in the forms of powder, granules, or clumps. In this drying process, the products are not placed in drying cabinets or shelves, but dispersed as fine droplets suspended in the air inside the dryer. The advantages of this method are that the technology is well known thus easily obtained; it can be used to produce capsules in large quantities, the coating materials for spray drying are approved as food products, and the coating materials dissolve in water and can release the core without leaving residue. Efendi stated that microencapsulation with spray dryer should utilize encapsulant materials with high solubility, emulsion-forming capability, layer-forming capability, dry, and low viscosity [5]. Even though several encapsulants can be used in nonfood materials, those for food products are limited to natural gum, carbohydrates, maltodextrin, wax, and several proteins.

Drying with spray dryer is performed by spraying the materials to be dried as mists, which increases the surface area of the materials to be in contact with the drying medium, thus the water evaporation process can proceed well. The spraying process is influenced by the form of the sprayer, speed of product flow, and product characteristics [6].

The spray dryer process consists of four stages: (1) atomization, in which liquid or paste is converted into mists, (2) contact between the atomized materials with hot air, (3) water evaporation from the materials to reach the desired moister content, and (4) product collection in a powder form. In the stages of spray drying process, there are several operational units consisting of preconcentrated solution, atomization (mist formation), drying using dry and hot air, separation of powder from water vapor, cooling, and product packaging.

2. Microencapsulation process of *turmeric* (*Curcuma domestica Val.*)

Turmeric (*Curcuma domestica Val.*) is a type of rhizoma medicinal plant containing curcuminoids, which consist of curcumin compound and its derivatives, desomethoxycurcumin and bis-desomethoxycurcumin. Curcuminoid is an active compound from *turmeric* rhizome that has biological activities with a wide application such as antihepatotoxic [7]. *Turmeric* has been known and used by the wider public, in the urban and rural areas, especially at homes, because of its wide usage. Part of *turmeric* used is the roots or rhizome, which is frequently used as organic fabric coloring, food flavoring, spices, and cosmetic materials. *Turmeric* is also used as traditional medicine for itching, gum inflammation, wounds, breathing shortness, stomachache, boils, skin fungal infection, back pain, jaundice, bad digestion, diarrhea, toxin neutralizer, low appetite, and so on [8].

Microencapsulation is a coating technology for solid, liquid, and gas using capsules in minute form, in which those capsules can release the core under specific conditions. Microencapsulation aims to protect sensitive components, reduce nutrient loss, and add food products in liquid form to solid form for ease of handling [9].

In this study, the microencapsulation process uses spray drying, which is the most frequently employed in the food industry because of its relatively lower cost. The advantages of this process are flexible and can be used for a variety of materials in microencapsulation because the equipment can be applied to process various materials and produce good quality particles with a consistent distribution of particle size. The food materials that can be applied in this method include fats, oils, and flavor enhancers. The coating can be from carbohydrates, such as dextrin, sugar, starch, and gum, or proteins, such as gelatin and soy proteins. Microencapsulation process includes emulsion formation or suspension on the active compounds and coating, and atomization of the emulsion into circulated dry and hot air inside drying chamber using an atomizer or a nozzle. The water contents inside emulsion droplets evaporate. The solid left over from the coating material traps the core material. Spray drying is useful for food materials that are sensitive to heat because the drying process occurs very fast. The other advantages of spray drying are the variety and availability of equipment, microcapsule quality that stays high, variety of particle size that can be produced, and good dispersibility in liquid media. However, loss still happens to active compounds with low boiling point. Physical characteristics of microcapsules depend on hot air (about 150–200°C), degree and uniformity during emulsion atomization, degree of emulsion density (30–70%), and emulsion temperature. The other disadvantages are the loss of bioactive compounds with low boiling point, oxidation in flavor enhancer substances, and limited options for shell materials, in which these materials can dissolve in water in an adequate amount. The flow diagram for microencapsulation process for *turmeric* to produce *turmeric* powder is presented in **Figure 1**.

This study was conducted to determine the optimal temperature of the inlet (T*inlet*) drying of the spray dryer to produce *turmeric* powder. *Turmeric* concentrate at 300 ml is added with 10% maltodextrin as microencapsulant; then, it was homogenized. The sample was homogenized using a magnetic stirrer to keep homogenized throughout the spray drying process. This process was taken into the spray dryer *SD-basic LabPlant*. Data were obtained from the same amount of volume but at different T*inlets* of 100, 120, 130, 140, 150°C, while the T*outlets* are recorded at 80–100°C, P (*blower*) at 4 m^3/mm, and feed flow at 0.6 ml/s. The yield of microencapsulation was calculated with definition reported by [10]. The microencapsulation yield was defined as percentage of total *turmeric* dried powder and the mass of the total *turmeric* liquid fed to spray dryer. The microencapsulation yield of *turmeric* powder drying at varied drying temperatures is given in **Table 1**.

In this study, the *turmeric* powder was produced by varying the T*inlet* on the spray dryer at five points of temperature, i.e., 100, 120, 130, 140, and 150°C. The samples contain 300 ml of *turmeric* concentrate and 10% maltodextrin [11].

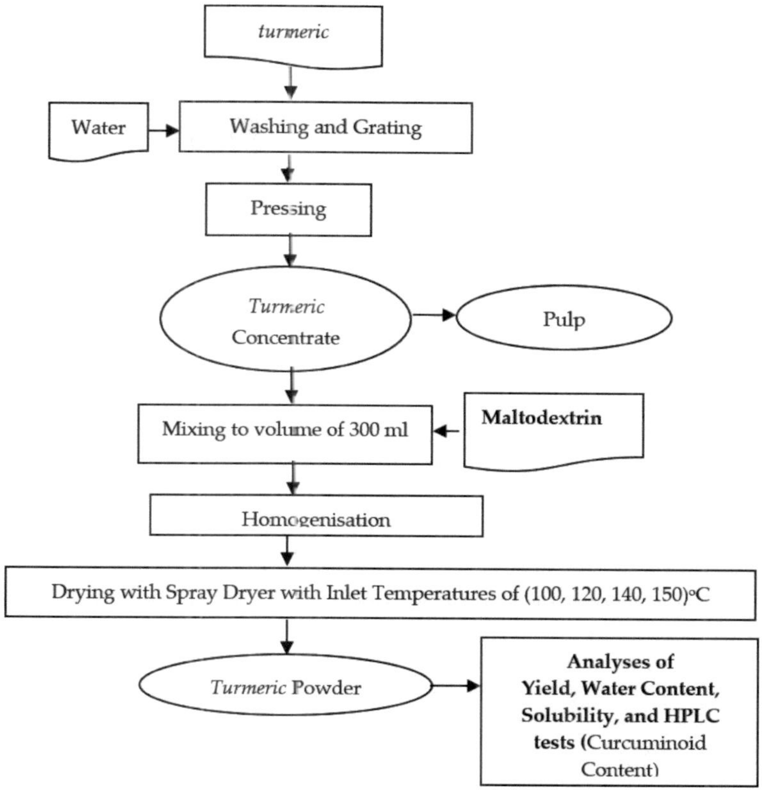

Figure 1.
Flow diagram of microencapsulation process transforming turmeric to turmeric powder [13].

No.	Samples for varied T*inlet* (b)	T*inlet spray dryer* (°C)	*Turmeric* concentrate (ml)	Drying time (s)	Yield (w/v) (%)
1	Sample I	110	300	480	2.64
2	Sample II	120	300	475	3.05
3	Sample III	130	300	478	3.31
4	Sample IV	140	300	460	3.69
5	Sample V	150	300	455	4.32

Table 1.
The study results of turmeric *powder drying time and yields at varied* Tinlet *on the spray dryer, using 10% maltodextrin.*

Based on the results shown in **Table 1**, the increased drying temperature (T*inlet*) reduces drying time, while increases yields. This study showed that at temperature of 150°C, the drying time took place at 455 s (7 min, 35 s) and produced 4.32% yield. The higher the T*inlet* spray dryer, the shorter the drying time. To obtain the optimum operation condition, it should observe also the curcuminoid content from **Table 4**.

This outcome is caused by higher the drying temperature, faster the water evaporation from the materials. The result of this study is supported by Estiasih et al. [6], where there is a difference of temperatures between heating medium and materials, in which the faster the heat transfers to the materials, the faster the water evaporates from them. As such, it can be understood that the higher the temperature used in the drying process, the shorter the drying time. However, it takes longer time for the spray dryer to reach higher temperatures.

The *turmeric* powder resulting from the spray drying process is tested for water content, solubility, and yield. The results of these tests are presented in **Table 2**. From this table, it can be seen that the water content, solubility, and yields of *turmeric* powder are affected by T*inlet* on the spray dryer equipment.

Water content analyses are performed to determine the water content of the powder produced from the spray dryer because water content influences shelf life, appearance, and water solubility. An increase of drying temperatures will reduce water content in the product. Water content testing is a part of quality testing on the *turmeric* powder and is conducted by heating at 105°C for 3 h, as described in SNI 01-2891-1992 on testing of food and beverage. The results of water content testing are presented in **Table 3**.

Based on **Figure 2**, an increase of drying temperature would reduce the water content of the product. This is because drying temperature has a role in water evaporation from the materials. And thus, the higher the temperature, the more water will evaporate, and the less water is left in the product.

Solubility is an important factor in powder product testing. Powder solubility is determined by composition, conditions during drying process, solvent temperatures, and mixing method. The higher the drying temperature, the less the water content in the products. The solubility testing is conducted by dissolving the *turmeric* powder samples produced at different T*inlets* in water at 100°C and recording the dissolving time in seconds. The effect of different T*inlets* of the spray dryer on the solubility of *turmeric* powder is presented in **Figure 3**.

As shown in **Figure 3**, the yields from a drying process are determined by the amount of the resulting products. In this study, the yields range from 1 to 4.42%, which means that the yields are relatively low compared to the initial dry materials

No.	Samples from different T*inlet* (b)	T*inlet* spray dryer (°C)	Water content (w/v) (%)	Dissolving time (s)
1	Sample I	100	8.5	492
2	Sample II	120	5.85	497
3	Sample III	130	4.15	520
4	Sample IV	140	4	532
5	Sample V	150	2.65	592

Table 2.
Test results on water contents and solubility of turmeric *powder produced at different temperatures of* Tinlet *on the spray dryer, using 10% maltodextrin.*

No.	T*inlet* spray dryer (°C)	*Turmeric* powder water content (%w/v)
1	110	8.5
2	120	5.85
3	130	4.15
4	140	4
5	150	2.65

Table 3.
Results of water content testing on turmeric *powder produced at different drying temperatures.*

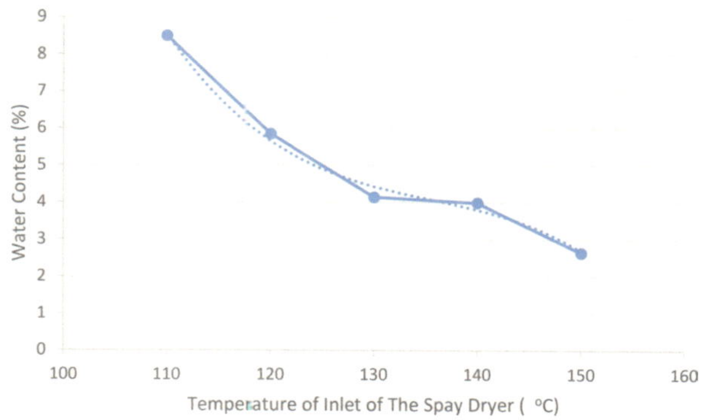

Figure 2.
The effects of temperature of inlet of spray dryer or drying temperature on the water content of turmeric *powder.*

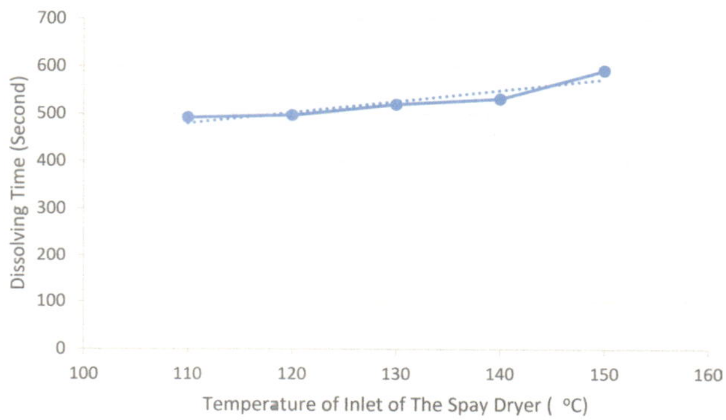

Figure 3.
The effects of temperature of inlet of spray dryer or drying temperature on the solubility of turmeric *powder.*

that are inserted in the spray dryer in liquid forms. In a drying process, free water molecules on the surface of the material particles can be easily evaporated, which produce low yields. However, based on the drying temperature variables, as presented in **Table 4**, the higher the drying temperature, the higher is the yield. It can be noted that the highest yield is found at the temperature of 150°C, and the effect of different T*inlet* on yields is presented in **Figure 4**.

Based on **Figure 4**, the effects of drying temperatures can be explained by an increase of temperatures causing dryer particles, which leads to less materials sticking inside the dryer and more getting collected in the cyclone vacuum collector. With an increase of temperatures, the yields obtained increase, and in this study, the highest yield is obtained from 150°C drying temperature at 4%. At the drying temperature of 100°C, the yield is relatively low at only 2.64%. The results of this study show that the drying temperatures have a positive correlation with the yields, such that when temperature is raised up to 150°C, the yields also increase because more materials are collected in the cyclone vacuum collector.

The results from HPLC testing are used to show curcuminoid contents in the *turmeric* powder samples and are presented in **Table 4**.

From the results of the curcuminoid content testing, it can be observed that an increase of drying temperature produces lower amount of curcuminoid contents, which is caused by the inability of curcuminoid compounds to be preserved by maltodextrin, as the microencapsulant. The best temperature to preserve curcuminoid compounds is at 110°C, in which 10.52% is preserved, although the yield was lower and drying timer was longer than 150°C.

Sample code	T*inlet* of spray dryer (°C)	Bis-demethoxycurcumin content (%)	Demethoxycurcumin content (%)	Curcumin content (%)	Total curcuminoid content (%)
Sample I	110	0.32	2.53	7.67	10.52
Sample II	120	0.22	1.20	3.63	5.05
Sample III [*13]	130	0.09	0.70	2.22	3.01
Sample IV	140	0.08	0.64	2.04	2.75
Sample V	150	0.07	0.29	1.29	1.65

Table 4.
*The results of HPLC on the curcuminoid contents *[13].*

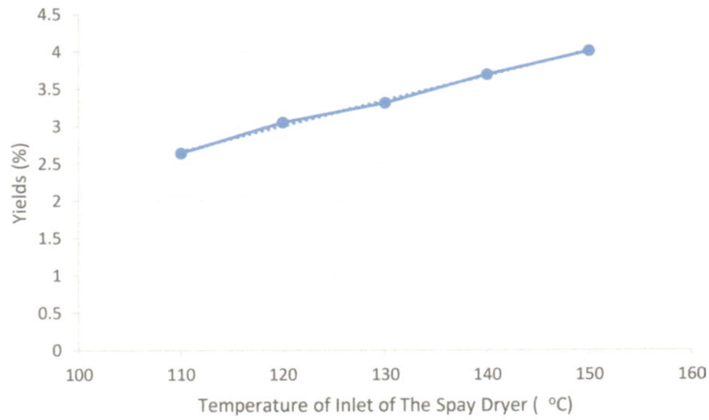

Figure 4.
The effects of temperature of inlet of spray dryer or drying temperature on the yields of turmeric *powder.*

Based on research results, *turmeric* contains many chemical substances that are useful for human body. Several chemical contents from *turmeric* rhizome that have been identified are essential oils at 6% consisting of monoterpenes and sesquiterpenes (zingiberene, alpha- and beta-turmerones), yellow coloring call curcuminoid at 5% (consisting of curcumin 50–60%, mono-desmethoxycurcumin, and bi-desmethoxycurcumin), proteins, phosphorus, potassium, iron, and vitamin C. Out of those three curcuminoid compounds, curcumin makes up the largest amount and one of its functions is to increase appetite in children (**Figures 5** and **6**).

The following figure shows the resulting chromatograms from the HPLC testings on sample V.

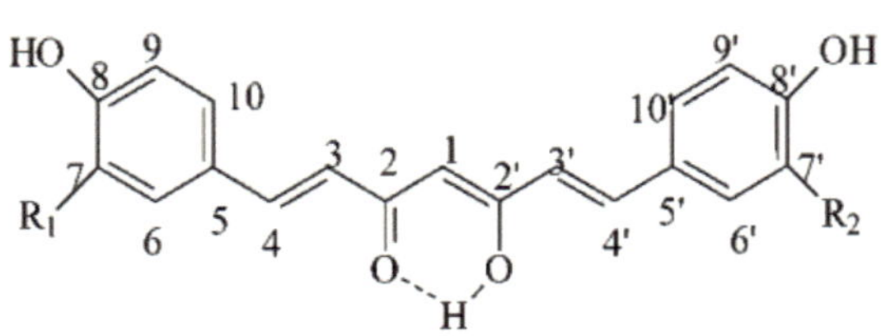

Compound	R1	R2
Curcumin (1)	OMe	OMe
Demethoxycurcumin (2)	H	OMe
Bisdemethoxycurcumin (3)	H	H

Figure 5.
Chemical structure of curcuminoid [12, 23].

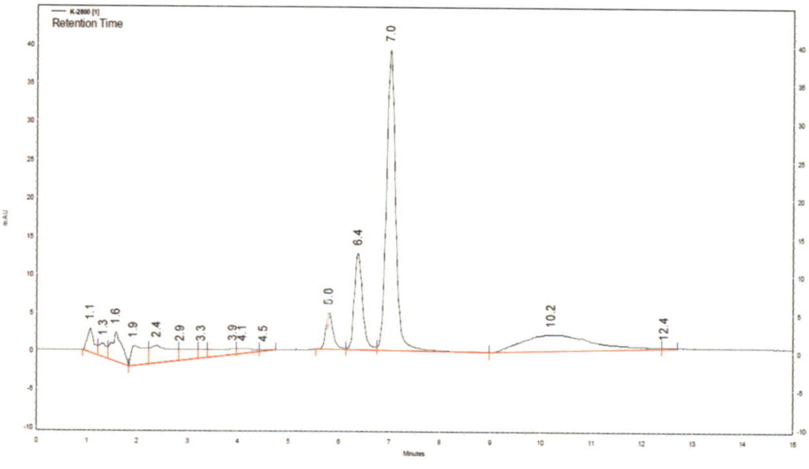

Figure 6.
Chromatograms of curcumin powder on sample I.

3. Microencapsulation process on *Aloe vera* (*Aloe chinensis Baker*)

Aloe vera plant is categorized as a low shrub, with succulent characteristics, and suitable for dry regions. The stem is short with the leaves forming a rosette around the stem and bell-shaped flowers. The leaves, which are the main parts to be utilized, have the lengths of 40–90 cm, widths of 6–13 cm, and thickness of 2.5 cm at the base. The variety that is generally cultivated in Asia, including Indonesia, is *Aloe chinensis* Baker, as described by Baker in 1977, which was developed in but is not native to China. This variety has been commercially grown in Indonesia, especially in the province of Kalimantan Barat (West Kalimantan) and more locally known as *Aloe vera* Pontianak [15–17].

The highly perishable nature of *Aloe vera* gel has prompted some efforts to process the gel harvest into powder. The aims are not only to preserve the contents of the gel, but also to increase the value of the harvest. So, *Aloe vera* is not just sold in fresh leaves, which usually are priced relatively low [15].

Furnawanthi [15] stated that *Aloe vera* in powder form also has other advantages: preserved nutrient contents, longer shelf life, and efficient transport. The raw material to powder ratio is around 150:1, which means that to obtain 1 kg of powder, 150 kg of fresh leaves are needed. As such, the establishment of *Aloe vera* powder industry requires a large amount of raw materials. This industry can also minimize the possibility of detrimental price drops that are often caused by overproduction and storage limitations of *Aloe vera* farmers. *Aloe vera* has high water contents and appears to be challenging to convert into powder. However, considering the contents of beneficial active compounds, several milling methods have been conducted to obtain those active compounds. The milling or drying technique frequently used is spray drying, whereas the common method is microencapsulation. Production of *Aloe vera* powder consists of two stages, which are (1) production of cores for *Aloe vera* powder and (2) drying with spray dryer. *Aloe vera* gel is crushed, blended, filtered, and vacuum evaporated to produce *Aloe vera* powder core. Microencapsulation uses maltodextrin as microencapsulant in a spray drying process. *Aloe vera* processing produces wastes in the form of rinds/pulp in a large amount. *Aloe vera* rinds are rich in organic materials or cellulose or pectic, and they can cause pollution problems if not managed. One of the waste managements is to use the by-products to make *Aloe vera* tea, livestock feed, and organic/composted fertilizer that is eco-friendly.

The procedures to produce *Aloe vera* powder were the following: (1) the *Aloe vera* was peeled and taken the gel, manually using knife; (2) the *Aloe vera* gel was crushed using blender. Then, it was filtrated using manual filter press, the filtrate was collected, and the pulp was thrown away; (3) the filtrate of *Aloe vera* was evaporated (40 times) using rotary vacuum evaporation (volume 8 lt) to get core of gel at temperature 35–40°C and vacuum condition (75–100 mbar); (4) the core of *Aloe vera* taken from evaporation was mixed with maltodextrin as filler and then it was mixed well using homogenizer with 1:1 composition between the core and maltodextrin. Then, it was homogenized until its concentration 50 °Brix (40–60 °Brix); and (5) In this research [17], the drying was conducted using spray drier. The hot air was introduced cocurrent with feed stream. In this stage, it was obtained the optimum variable process for drying to get active compound still maintained. The optimization was conducted to obtain the optimum drying temperature corresponding to desired quality of product or product in the market. To approach this, the drying temperature was varied: 110, 120, 130, and 140°C [17]. Mass flow diagram of *Aloe vera* powder production from the initial mass of 100 kg of *Aloe vera* leaves is presented in **Figure 7**.

To obtain the optimum drying temperature, the optimization was conducted to preserve the active compounds corresponding to commercial *Aloe vera* powder. The optimization was carried out at 110, 120, 130, and 140°C drying temperatures. The result shows that the density was almost same respectively to the commercial products (Terry Labs' product). Hence, the water content was below the commercial product. The water content was 2.88–4.89% w/w in which the commercial product is 8% max. This might be because of drying process. In the cocurrent spray dryer, the hot air is contacted with the feed in the same stream; it means that the highest temperature of hot air meets with the first feed stream. The microencapsulated active components have been affected by high temperature of dryer; on the other hand, it was relatively stable at lower temperature. It means that the quality of product has been affected by temperature of dryer. To analyze the chromatography result, an LC-MS method was conducted at absorbance of 254 mm for *Aloe vera* gel powder concentrates achieved from evaporation process, and the *Aloe vera* powder after 110, 120, 130, and 140°C drying temperatures. The result shows that *Aloin* A and B, *Aloenin* (B), *aloesin*, and *Chrysophanol* were appeared in all samples. Although *Aloe-emodin* was not detected in all samples, *Aloeresin* A was appeared in evaporated *Aloe vera* gel and *Aloe vera* powder (110 and 120°C drying temperature). Based on the result, the optimum drying temperature was 120°C to produce *Aloe vera* powder where all of phenolic compounds of *Aloe vera* powder was still maintained [17].

The *Aloe vera* powder from fresh *Aloe vera* leaves was analyzed for the microbiology, water content, density, solubility, pH, particle, color, and active component using LC-MS. The properties of *Aloe vera* powder obtained from the research for dryer temperature variation were described in **Table 5**. It was compared with the standard commercial *Aloe vera* powder from Terry Labs.

In general, the resulting product has met most of the parameters and specifications of commercial *Aloe vera* powder on the market such as water content, solubility, color, pH, appearance, and microbiology. **Table 5** shows that drying with higher temperatures resulting in *Aloe vera* powder products with microorganism contamination levels is lower even though the four variables still eligible.

While the product is almost the same density compared to available commercial products, this might be due to the method of testing using different methods, so the result is somewhat different. The testing methods used packed density. In the drying process (spray dryer), the decreasing of hot air inlet temperature did not affect the increase of water content significantly. In fact, water content tended

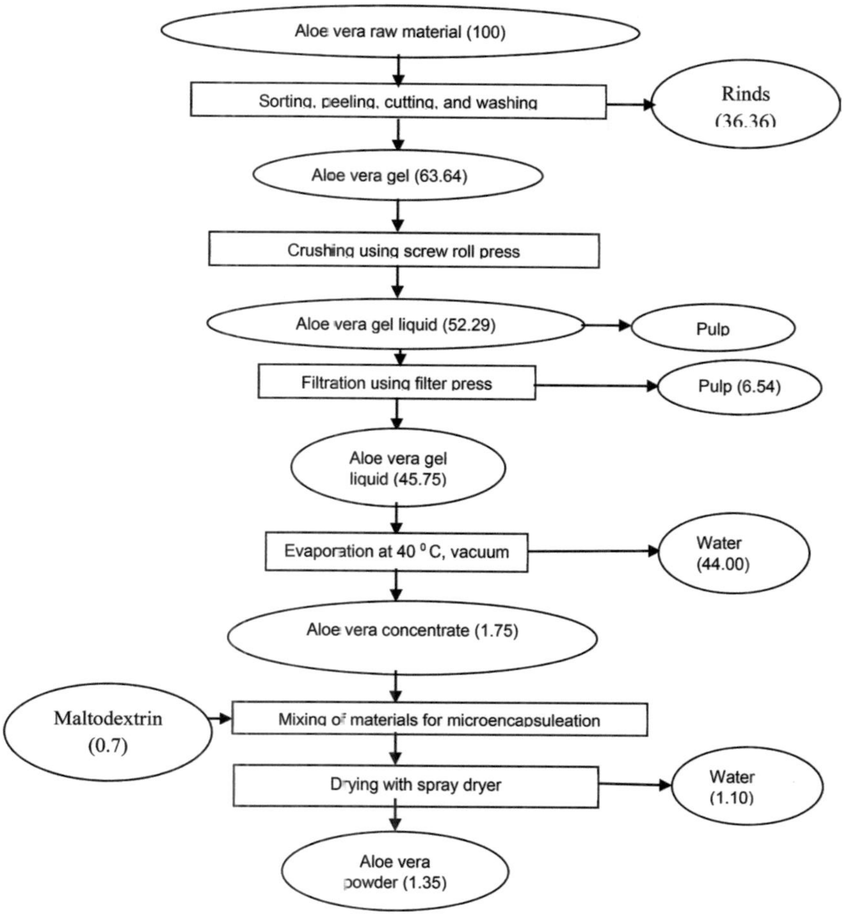

Figure 7.
Mass flow diagram of Aloe vera powder production from the initial mass of 100 kg of Aloe vera leaves.

No.	Compounds	*Aloe vera* powder 1 (140°C)	*Aloe vera* powder 2 (130°C)	*Aloe vera* powder 3 (120°C)	*Aloe vera* powder 4 (110°C)	*Aloe vera* powder spray-dried gel (Terry Labs' product)
1	Water content (% w/w)	2.88	4.04	4.89	4.89	8% max
2	pH	4.98	4.99	4.97	4.98	3.5–5.0
3	Microbiology (cfu/g)	96	97	97	98	<100
4	Density (g/ml)	0.99	0.99	1.00	1.00	0.990–1010
5	Solubility (min)	2.26	1.93	2.94	2.94	5
6	Color	Beige white	Beige white	Beige white	Beige white	Beige white
7	Appearance	Fine	Fine	Fine	Fine	*Fine crystalline powder*

Table 5.
*The properties of **Aloe vera** powder obtained [17].*

No	Specifications	Spray-dried aloe gel powder	No	Specifications (amino acid contents)	Spray-dried aloe gel powder, ppm
1	Amylase activities	0.024 unit/gr sample	1	Aspartic acid	131.71
2	Cellulose	0.0197%	2	Glutamic acid	153.12
3	Lignin	0.0089%	3	Serine	88.25
4	Saponin	Confirmed presence (qualitative test)	4	Glycine	72.78
5	Glucose	48.45 ppm	5	Histidine	155.23
6	Calcium (Ca)	0.93%	6	Arginine	135.92
7	Magnesium (Mg)	0.13%	7	Threonine	155.93
8	Phosphor	37.3 ppm	8	Alanine*	65.94
9	Lead (Pb)	<0.02 ppm	9	Proline	132.11
10	Arsenic (As_2O_3)	<0.005 ppm	10	Tyrosine	242.98
11	Zn	0.05%	11	Valine*	127.39
12	Natrium (Na)	0.73%	12	Methionine*	192.79
13	Kalium (K)	0.51%	13	Cysteine	106.29
			14	Isoleucine*	223.26
			15	Leucine*	166.01
			16	Phenylalanine*	124.08
			17	Lysine*	174.24

**Essential amino acid.*

Table 6.
The results of chemical analyses on the contents of Spray-dried aloe gel powder [14, 15].

to be stable of 2–5%. This has a positive effect for the quality of product in which the active component microencapsulated was relatively stable for lower temperature of dryer.

The results of chemical and content analyses of active compounds in *Aloe vera* powder are presented in **Table 6**. From these results, it can be determined that *Aloe vera* powder can be used in cosmetics, pharmaceutical, and food industries. In these industries, the functions of these bioactive compounds must be preserved. The lignin and saponin contents make *Aloe vera* powder very suitable for skin care formulations, such as lotion, wash, shampoo, and soap. The contents of active compounds in *Aloe vera* powder are complete with proteins, polysaccharides, lignin, saponin, and minerals, and can be incorporated into formulations for topical applications, such as anti-plaque toothpaste, shampoo, soap, lotion, sunscreen, and burn cream; whereas for internal uses, *Aloe vera* powder can be used as diabetic medication, because of its high polysaccharide content, and dietary and health supplements [18–23].

4. Conclusion

Microencapsulation proposes to protect sensitive food components, reduce nutritional losses, expand the usefulness of sensitive food components, add certain food to other food, protect flavors and fragrances, convert liquid food components to more convenient solids handled and protected materials from environmental influences. Product microcapsulation can be used as raw material for the food industry, cosmetics, and pharmaceuticals, using bioactive compounds. From the results of the curcuminoid content testings, it can be observed that an increase of drying temperature produces lower amount of curcuminoid contents, which is caused by the inability of curcuminoid compounds to be preserved by maltodextrin, as the microencapsulant. The best temperature to preserve curcuminoid compounds is at 110°C, in which 10.52% is preserved. Hence, for *Aloe vera* processing, the optimum drying temperature was 120°C which maintained the active component of *Aloe vera* powder. The result of LC-MS observed that the active components of *Aloe vera* powder can be maintained at the optimum operation condition of drying. The optimum drying temperature was 120°C, which was the active component of *Aloe vera* powder such as *Aloenin (B)*, *Aloeresin A*, and *Chrysophanol* still maintained.

Acknowledgements

Thanks to Universitas Muhammadiyah Jakarta and Directorate of Research and Community Service, Directorate General for Research and Development, Ministry of Research and Technology Higher Education on the Research Grant PTUPT in 2018, Contract number 006/KM/PNT/2018, 06 March 2018. Thanks to Rector of Universitas Muhammadiyah Jakarta and Dean of Engineering Faculty, Universitas Muhammadiyah Jakarta.

Author details

Tri Yuni Hendrawati[1*], Alvika Meta Sari[1], Muhamad Iqbal Syauqi Rahman[1], Ratri Ariatmi Nugrahani[1] and Agung Siswahyu[3]

1 Chemical Engineering Department, Engineering Faculty, Universitas Muhammadiyah Jakarta, Indonesia

2 Chemical Engineering Department, Industrial Engineering Faculty, ISTA Jakarta, Indonesia

3 Chemical Engineering Department, Al Kamal Science and Technology Institute, West Jakarta, Indonesia

*Address all correspondence to: yuni.hendrawati@ftumj.ac.id

References

[1] Bertolini AC, Siani AC, Grosso CRF. Stability of monoterpenes encapsukated in gum aeabic by spray drying. Journal of Agricultural and Food Chemistry. 2001;**49**:780-785

[2] Franjione J, Niraj V. The Art and Science of Miecroencapsuation. New York: Botanical Garden Press; 2003

[3] Risch SJ. Encapsulation: Overview of uses and techniques, di dalam. In: Risch SJ, Renescius GA, editors. Encapsulation and Controlled Release of Food Ingridients. Washington.D.C: American Chemical Society; 1995

[4] Hogan SA, Namme M, Riordan EP, O'Sullivan M. Microencapsulating properties of sodium caseinate. Journal of Agricultural and Food Chemistry. 2001;**49**:1934-1938

[5] Efendi E. Mikroenkapsulasi Minyak Atsiri Jahe dengan Campuran Gum Arab–Maltodekstrin and Variasi Suhu Enlet Spray dryer [Tesis]. Yogyakarta: Program Pasca Sarjana. UGM; 2000

[6] Estiasih T. Mikroenkapsulasi Konsentrat Asam Lemak N-3 dari Limbah Cair Pengalengan Ikan Lemuru (*Sardinella longiceps*). [Tesis]. Yogyakarta: Program Pasca Sarjana. UGM; 1996

[7] Sujatno M. Efek attapulgit, ekstrak daun psidium guajava, dan ekstrak akar curcuma domestica terhadap diare akut nonspesifik. Majalah Kedokteran Indonesia. 1997;**46**(4):199-200

[8] Rukmana R. Turmeric. Yogyakarta: Kanisius; 1999. Cetakan pertama

[9] Dziezak JD. Microencapsulation and Encapsulation Ingridients. Food Technology. 1988;**2**(4)

[10] Sutrisno K. Teknologi enkapsulasi flavor rempah-rempah, Swapnali. 2005

[11] Susantikarn P, Donlao N. Optimization of green tea extracts spray drying as affected by temperature and maltodextrin content. International Food Research Journal. 2016;**23**(3):1327-1331

[12] Van der Goot H. The chemistry and qualitative structure-activity relationship of curcumin. In: Pramono S, Jenie UA, Retno SS, Didik G, editors. Proceedings of the International Symposium on Curcumin Pharmacochemistry (ISCP). Yogyakarta: Faculty of Pharmacy Gadjah Mada University; 1997. pp. 13-27

[13] Hendrawati TY, Mubarok MA, Ramadhan AI. The effect comparison Maltodextrin against results characteristics of Microencapsulation of turmeric (curcuma Domestica Val). ARPN Journal of Engineering and Applied Sciences. 2017;**12**(13):4129-4135

[14] Jayaprakasha GK, Jagan Mohan Rao L, Sakariah KK. Improved HPLC Method for the Determination of Curcumin, Demethoxycurcumin, and Bisdemethoxycurcumin. Journal of Agricultural and Food Chemistry. 2002;**50**:3668-3672

[15] Furnawanthi I. Khasiat dan Manfaat Lidah Buaya Si Tanaman Ajaib, PT. Jakarta: Agromedia Pustaka; 2003

[16] TY Hendrawati M, Eriyatno ISK, Sunarti TC. Rancang bangun industri tepung lidah buaya (aloe vera) terpadu. Journal of Agroindustrial Technology. 2007;**17**(1):12-22

[17] Hendrawati TY. Aloe Vera powder properties produced from aloe Chinensis baker, Pontianak, Indonesia. Journal of Engineering Science and Technology. Special Issue on SOMCHE 2014 & RSCE 2014 Conference, January (2015). School of Engineering, Taylor's University. 2015:47-59

[18] Changa XL, Wanga C, Fengb Y dan Liua Z. Effect of heat treatment on the stabilities of polysaccharides substances and barbaloin in juice from aloe vera miller. Carbohydrate Research. 2006;**341**(3):355-364

[19] Chowa JTN, Williamson DA, Kenneth M, dan Gouxa WJ. Chemical charaterization of the immunomodulating polysaccaharide of Aloe vera L. Journal of Pharmaceutical and Biomedical Analysis. 2005;**37**(5):937-941

[20] Elamthuruthya AT, Shahb CR, Khanb TA, Tatkeb PA, dan Gabheb Y. Standarization of marketed kumariasava an ayurvedic aloe vera product. Food Control. 2004;**16**(2):95-104

[21] Eshun K dan He Q. Aloe vera: A valuable ingredient for food, pharmaceutical and cosmetic industries. International Journal of Aromatherapy. 2004;**14**(1):15-21

[22] Morsy EM. Aloe Vera Stabilization and Processing for the Cosmetic Beverage and Food Industries. 5th ed. USA: Citra International; 1991

[23] Wu JH, Xu C, Shan CY d, Tan RT. Antioxidant properties and PC12 cell protective effect of APS-1, a polysaccharide from aloe vera var. Chinensis. 2006;**39**(1):93-100. Postharvest Biology and Technology

Section 3

Examples of Applications

Chapter 5

Applications of Microcapsules in Self-Healing Polymeric Materials

Seng Neon Gan and Nurshafiza Shahabudin

Abstract

Self-healing polymeric materials have a great potential to be explored and utilized in many applications such as engineering and surface coating. Various smart materials with self-healing ability and unique self-healing mechanisms have been reported in recent publications. Currently, the most widely employed technique is by embedding microcapsules that contain a healing agent into the bulk polymer matrix. When cracks develop in the polymer matrix, the curing agent is released from the microcapsules to cross-link and repair the cracks. Microencapsulation of the healing agent in the core can be achieved by *in situ* polymerizing of shell material. This chapter presents a general review on self-healing materials, and particularly, self-healing of epoxy matrices that includes epoxy composite and epoxy coating by microencapsulation technique. Microencapsulation processes, including types of resin used, processing parameters such as core/shell ratio, concentration of emulsifiers, viscosities of aqueous and organic phases and stirring rate are discussed.

Keywords: self-healing, epoxy, microencapsulation, microcapsule

1. Microcapsule-based self-healing epoxy composite and epoxy coating

The use of microcapsule-based systems has attracted much attention during the last decade. The advantages include the esthetic recovery combined with the fast release of healing agent from the microcapsules [1]. This microcapsule-based approach could be easily integrated in many polymer systems, although the healing agent is locally depleted after a single damage event [2]. The attractive features of such technology include prolonging the service life-span of the materials and reducing the cost of repair or replacement of failed component.

Designing a microcapsule-based self-healing material requires a number of considerations. The first is to design the microcapsules, followed by incorporating them into the polymer matrices. Next, the viable mechanical characterization of the polymer during the occurrence of cracks should be studied. Finally, the extent of self-healing reactions must be determined and verified. Thus, the suitable encapsulation method for a particular healing agent must first be identified. One must consider the operation parameters such as solubility, reactivity, viscosity, and volatility of the healing agent. In the next step, the microcapsules should be integrated into the matrix host without being broken during mixing and must be well distributed. Microcapsules produced by encapsulating the healing agent in urea-formaldehyde (UF), melamine-formaldehyde/melamine-urea-formaldehyde (MF/MUF), and poly(urethane) (PU) microcapsules have been shown as being capable of withstanding processing conditions in thermoset resins as well as in composite materials. The mechanical properties, triggering mechanism, and self-healing performance of the material should be characterized to ensure the usefulness of the innovation. The overall properties of the healed materials, such as fracture toughness, tensile strength, and hardness could be affected by the microcapsules [2].

1.1 Types of microcapsules for self-healing

Different types of microcapsule systems have been applied in self-healing materials; these include the microcapsule-catalyst system, dual/multi-capsule system, and microcapsules with latent hardener. Grubb's catalyst was first reported in a self-healing material containing micro-encapsulated dicyclopentadiene (DCPD) (White et al., [3]). Later, a self-healing epoxy adhesive was designed by incorporating 15 wt. % micro-encapsulated DCPD and 2.5 wt. % Grubbs' catalyst. A significant increase in the initial fracture toughness after 24 h of healing at ambient temperature was observed [4]. Healing process at elevated temperature of 110°C was applied in a rubber-toughened epoxy adhesive. To withstand specific epoxy hash curing conditions, the DCPD was encapsulated in a double-walled polyurethane (PU)/UF shell. Recovery of 20–58% of virgin fracture toughness was obtained after assessment with width-tapered double cantilever beam (WTDCB) test [5].

The microcapsules and catalyst self-healing system was found to have certain shortcomings. For instance, poor dispersion of the catalyst and the hardener (diethylenetriamine, DETA) could lead to a drop in healing efficiency. Thus, an alternative method of encapsulating the catalyst was introduced [6], by addition of wax microspheres into the epoxy matrix. The wax-catalyst microspheres were found to improve the dispersion, which resulted in efficient healing (up to a maximum of 93%) with significantly lower amount of embedded catalyst than the non-wax-encapsulated catalyst. Although the Grubbs' catalyst has excellent selectivity, it is costly and toxic. Therefore, these drawbacks have limited its use in high-volume commercial composite and polymeric parts.

Tungsten (VI) chloride was used as an alternative catalyst for DCPD; it is more cost-effective, widely available, and more tolerant to moderate temperature changes. It was used for catalyzing the ring-opening metathesis polymerization of *exo*-DCPD and achieved an *in situ* healing efficiency of approximately 20% with 15 wt. % microcapsules [7].

A different type of chemistry was involved when metal triflates were used as Lewis acid catalyst, to initiate the curing of diglycidyl ether of bisphenol A (DGEBA) and epoxy resins. It was easily available and thus chosen due to its high catalytic activity, and of relatively low cost, low toxicity, and high stability [8]. The DGEBA, ethyl phenyl acetate microcapsules, and scandium(III) triflate catalyst particles were embedded into epoxy matrix. The achieved healing performances were comparable to the more expensive and less robust (air and moisture-sensitive) Grubbs' catalyst/DCPD-capsule system [9].

Multi-capsule self-healing systems have also been reported. For example, a two-part resin system containing microcapsules of an epoxy resin and its hardener with poly(melamine-formaldehyde) (PMF) shell was employed. The PMF shell was inert toward the two types of cores. It was reported that at 20°C for 24 h of reaction time, 43.5% healing efficiency was achieved at 1 wt. % capsule content, and 100% healing efficiency at 5 wt. % capsule content [10]. Pang and Bong have used a similar approach earlier but with the healing agent loaded inside hollow fibers instead of microcapsules [11].

In another example of two-component microcapsule system, epoxy and boron trifluoride diethyl etherate ($(C_2H_5)_2O{\cdot}BF_3$) as hardener were used in self-healing epoxy composites. Boron trifluoride diethyl etherate has been commercially available as a cationic catalyst for low-temperature chain polymerization of epoxy resins. A remarkable recovery of 80% of impact strength was achieved within 30 min at 20°C at only 5 wt. % epoxy and 1 wt. % ($(C_2H_5)_2O{\cdot}BF_3$)-loaded microcapsules. The only shortcoming of this system was that the high reactivity of the healing agents resulted in limited shelf life [12].

Another approach is to encapsulate the epoxy resin and amine hardener separately without the need of another catalyst. The epoxy microcapsules were prepared by the classic *in situ* polymerization method, while the hardener-microcapsules were prepared by vacuum-infiltrating the amine into hollow PUF microcapsules. High healing efficiency with 91% recovery of mode-I fracture toughness was achieved with 7 and 10.5 wt. % amine and epoxy microcapsules, respectively. The microcapsules have shown 6-month storage stability at ambient conditions [13].

Recently, to overcome the challenges in encapsulating the hardener for epoxy resins, poly(methyl methacrylate) (PMMA) was used instead of the amino resins. The PMMA microcapsules can withstand 6–12 months of storage at room temperature [14]. They have obtained 43.5 and 84.5% fracture toughness recovery with 5 and 15 wt. % microcapsules, respectively, at room temperature for 24 h of curing [15].

The third type of self-healing system is the microcapsule-latent functionality where the healing agent is encapsulated or dispersed as particles, and the polymerizer is residual reactive functional groups in the matrix. This approach was introduced by Yin and colleagues. Encapsulated epoxy resin was used as curing agent and a well-dispersed latent hardener, which is also epoxy-based, was distributed in the matrix. Repair of the cracked sites was achieved through the cross-linking reactions of the released epoxy resin from the fractured microcapsules [16]. Another example of this system utilizes solvents together with the healing agent. Here, residual amine functionality in an epoxy matrix is used to initiate polymerization with healing agent [17]. Other works have incorporated meltable, thermally polymerizable epoxy microspheres into epoxy composite materials to induce self-healing [18] and also water-soluble, self-curing epoxy-amine adduct particles in a protective coating on a steel substrate [19].

The fourth type of self-healing system comprises a simplified processing method of capsule-catalysts. Here, the catalyst and the healing agent were dispersed throughout the matrix. Although encapsulated-catalyst healing system was used, the siloxane-based healing agents, that is, hydroxyl end-functionalized polydimethylsiloxane (HOPDMS) and polydiethoxysiloxane (PDES) mixture were not encapsulated. Due to their low solubility, the siloxane-based polymers and the encapsulated butyltin dilaurate catalyst mixture were directly blended with the vinyl ester prepolymer, forming a distribution of stable phase-separated droplets and protected catalyst (**Figure 1**). No reactions took place between the HOPDMS and PDES prior to exposure to the catalyst. A stable healing chemistry in humid or wet environments and a stable system for elevated temperature (important for higher temperature thermoset-curing system) was reported. The siloxane-based polymers are also widely available and comparatively low in cost [20].

1.2 Self-healing epoxy coating

Self-healing epoxy coating is another attractive application of the self-healing property. This system was first introduced when microvascular networks were included into epoxy coating. These networks then released the healing agent when cracks occurred, flowed to fill the gaps, and cross-linked to mend the damages [21]. Another work reported the healing of cracks in paint film by linseed oil released from microcapsules that ruptured under simulated mechanical action.

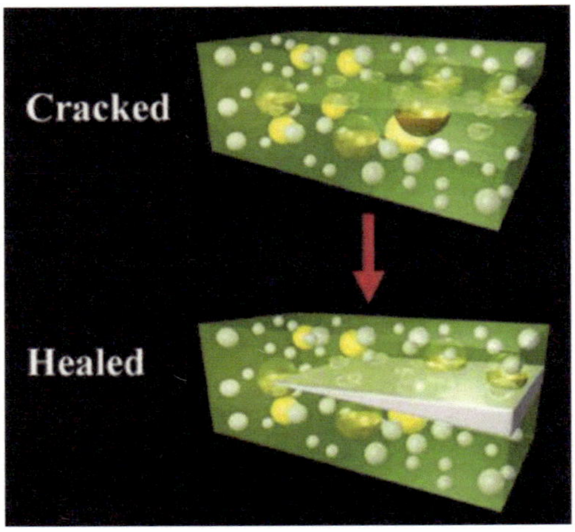

Figure 1.
Polydimethylsiloxane (PDMS)-based self-healing is achieved through the tin-catalyzed polycondensation of phase-separated droplets. Reproduced from Cho et al. ©2006 Wiley-VCH.

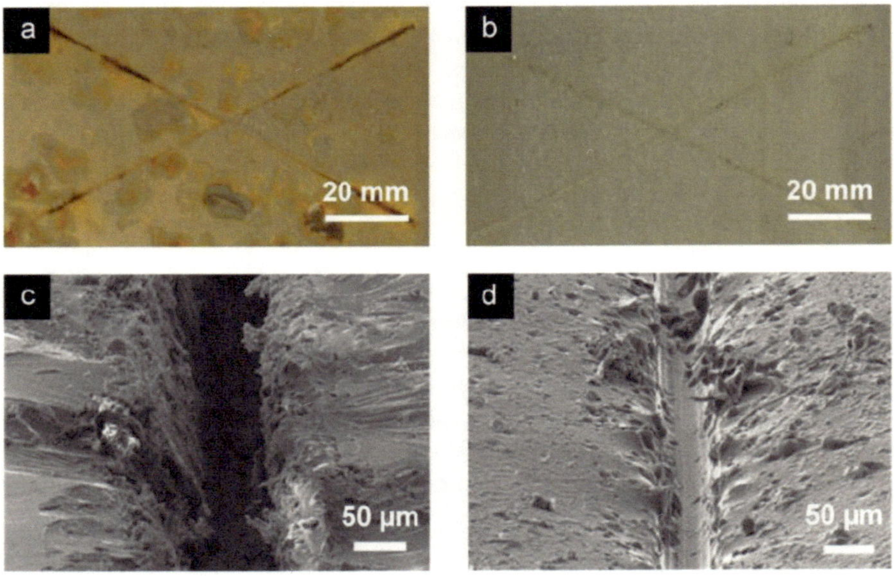

Figure 2.
Optical images after 120 h of immersion in salt water of (a) control sample (b) self-healing coating; SEM images of the scratched region of (c) the control coating (d) the self-healing coating after healing. Reproduced from Cho et al. © 2009 Wiley-VCH.

The healed area was able to protect the substrate from corrosion [22]. In another investigation, the encapsulated linseed oil along with solvents and dispersing agent were added to the epoxy coating. This was compared to another epoxy coating filled with talc at similar level. Results showed that the coating containing encapsulated linseed oil formed fewer and smaller cracks. The enhanced impact resistance of the microcapsule-filled coating might be due to higher elasticity of microcapsule containing linseed oil compared to the harder inorganic filler particles. Based on preliminary salt spray exposure of coatings, the addition of microcapsules to the epoxy binder matrix has not compromised the coating performance [23].

The possibilities of sonication to produce nanocapsules containing linseed oil as healing agent in epoxy coating was examined by Boura et al. [24] Capsules of smaller sizes provide easier dispersion for better healing performance for the coating matrix and had better wet adhesion and corrosion resistivity than those of larger sizes [24]. Tung oil was also encapsulated to impart self-healing ability in epoxy paint films. Scratches were healed efficiently with satisfactory anti-corrosive properties [25].

Encapsulated Grubbs' catalyst and DCPD in silica-coated micron- and sub-micron-size capsules had been demonstrated by Jackson and co-researchers [26]. Silica served as a protective and functional layer to the capsules and catalyst, and significantly improved dispersion in the epoxy matrix. The particles were successfully incorporated into the epoxy matrix without significant loss of healing agent. It also enabled the capsules and particles to be dispersed at higher concentrations with little loss of reactivity.

The concept of phase-separated self-healing was extended to an anti-corrosion epoxy coating. A dramatic reduction in corrosion of metal protected by the coating is shown by the optical images in **Figure 2** [27].

2. Microencapsulation of healing agent

Microencapsulation is the process of coating small solid particles, liquid droplets, or gas bubbles with a thin film of shell materials. The term microcapsule is used to describe particles with diameters between 1 and 1000 μm [28]. It consists of a core-shell structure where the active reagent is surrounded by a membrane (reservoir system) [29]. The shell of the microcapsules and the process of encapsulation are selected according to the physical properties of the core and the intended application.

Microcapsules containing self-healing agent are embedded into polymeric materials. When crack occurs in the polymer matrix due to an external force such as sudden physical impact or stretch, some of these microcapsules along the crack line are ruptured and the liquid curing agent flows into cracks and reacts with the catalyst or the matrix itself to form solid that fills the gaps. The healing agents could be monomers, dyes, catalysts, or hardeners, and they are encapsulated to suit the chemistry of the intended matrix and applications. These healing agents behave like adhesives that glue the polymeric matrix from inside the system. Ionomers, concrete, ceramics, metals, and polymers are among known materials that have been developed to have self-healing ability.

There are many methods to prepare the microcapsules such as interfacial polymerization, coacervation, *in situ* polymerization, extrusion, and sol-gel methods. Among these methods, the *in situ* polymerization was the most frequently used technique, because it is the easiest process, and does not require sophisticated equipment [30].

The shells of microcapsules were synthesized via *in situ* polymerization. The core material in the form of water-immiscible liquid (or solid) was dispersed in an aqueous phase that contains urea, melamine, water-soluble urea-formaldehyde condensate, or water-soluble urea-melamine condensate. At the same time, a suitable additive that could enhance shell formation is added to the aqueous phase if necessary. Shell formation begins to occur when formaldehyde is added and the pH adjusted to between 2 and 5 with a dilute mineral acid. The system is maintained at 40–60°C for several hours. Condensation reaction occurs in the aqueous phase to produce oligomers that deposit on the surface of the dispersed core particles and continue to polymerize to form a water-insoluble, cross-linked hard polymeric shell. The process has been used commercially to manufacture a wide range of microcapsules [28].

2.1 Resins for encapsulation

Amino resins are most widely used in making microcapsules for self-healing materials [30]. They are also known hardeners for epoxy resins. Their reactions occur first through the etherification process with loss of water or alcohol and second through the addition reaction of *N*-methylol groups to the epoxide group [31]. These resins are produced by the reactions of amino groups, thus they are compounds with -NH functionality such as urea, melamine, thiourea, aniline, and guanamines, which can react with formaldehyde. Urea-formaldehyde (UF) and melamine-formaldehyde (MF) are the two most popular amino resins, and their structures are shown in **Figure 3**.

Both UF and MF resins are light in color and exhibit enhanced water and heat resistance. They are widely used in industries as adhesives, impregnating resins, molding materials, strengtheners for building materials, flame-retardant coatings, foamed resins for many purposes, grinding wheels, ion-exchange resins, sewage flocculants, and microcapsule production [32, 33]. UF resin was also used in specialized application such as in the fabrication of natural fiber-reinforced plywood and chip-boards [34].

In preparing the microcapsules, UF can be cross-linked to form hard shells that protect the healing agent in the cores. Some UF nanoparticles would deposit on the shell to form a rough surface that facilitates the adhesion of microcapsules to the polymer matrix [35, 36]. During the preparation process, a low-molecular weight oligomer is first formed from the condensation of urea and formaldehyde. Subsequently, the oligomer becomes attached onto the surface of the dispersed core material and gradually polymerizes to form the shell [37].

A number of modifications of poly(urea formaldehyde) (PUF) resin for encapsulation has been reported, for example, by mixing urea with melamine-formaldehyde prepolymer forming poly(melamine-urea-formaldehyde) (PMUF). 5-ethylidene-2-norbornene (ENB) and its cross-linking agent were encapsulated for self-healing purpose [38]. PMUF shells are more robust, stronger, and easier to handle than PUF shells. Up to 12 wt. % urea was replaced with melamine in the formulation, to encapsulate an epoxy resin. The resultant PMUF microcapsules exhibited better resistance against solvent, acid, and alkali [39]. Other researchers had replaced 1–5% urea with a commercially available melamine resin, Cymel 303®, in producing shell materials for microcapsules containing DCPD. The microcapsules were strong enough to withstand the mechanical mixing with a viscous restorative dental resin [40]. Both PUF and PMUF resins have been used to prepare microcapsules of epoxy resins, DCPD, linseed oil, and alkylglycidyl ether. Here, it is

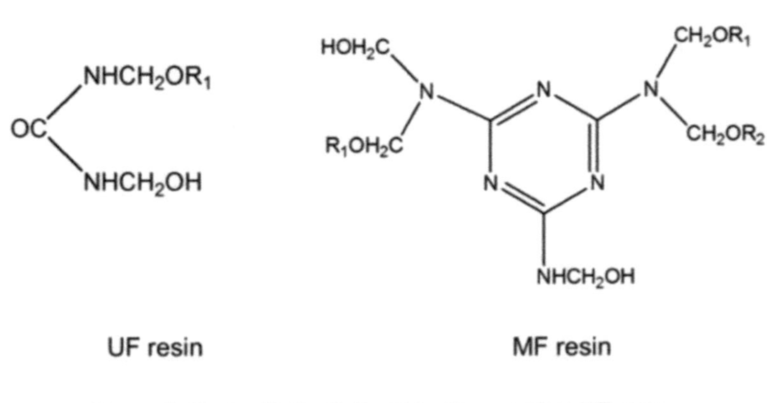

Figure 3.
Chemical structures of UF and MF resins.

also reported that PMUF shells were more stable than PUF shells and resulted in the production of higher yield of microcapsules [36].

2.2 Parameters affecting microencapsulation

There are several processing parameters that could affect the formation and the size of the microcapsules. These factors are discussed briefly as follows. The core/shell weight ratio was calculated based on the weight of core over the weight of the total raw materials forming the shell. Brown and his colleague have used a 6.2/1 ratio in the UF *in situ* polymerization [37]. This ratio has been adopted by a number of other researchers [39, 41, 42]. The tendency to form thicker and porous layers that cause agglomeration was noted by some researchers. A higher ratio of 6.45/1 was reported to produce more spherical and well-formed microcapsules [43].

The size of microcapsules was found to change with the weight ratio of core-shell material. The higher amount of core material results in larger size of microcapsules. Fixing the other processing parameters, an increase in the core/shell weight ratio increases the size of core droplet in emulsion. Consequently, the core material and the size of microcapsules increase. However, excess core materials may cause poorer dispersion, and there is relatively less oligomer to coat and fully polymerize to hard shell, resulting in more aggregation of core droplets, leading to lower yield of good microcapsules. The final microcapsules would have thinner shell wall and become fragile [44].

The concentration of emulsifiers has played a crucial role during *in situ* polymerization; if it is too low, the droplets will tend to agglomerate into bigger sizes, thus an increase in concentration may be required to maintain the sizes of the droplets within a certain range [45, 46]. Sodium dodecyl benzene sulfonate (SDBS) is one of the surfactants used in the production of microcapsules by *in situ* polymerization [44], while sodium dodecyl sulfate (SDS), gum Arabic, and gelatin are normally used in coacervation method [45, 47]. Poly(vinyl alcohol) (PVA), that serves to adjust the viscosity of the medium, is also used in PUF encapsulation [16, 22]. However, polyelectrolyte species are the most commonly used emulsifiers in the production of PUF microcapsules, such as ethylene maleic anhydride (EMA) copolymer, methylvinyl ether maleic anhydride copolymer, and styrene maleic anhydride copolymer [48].

The use of surfactants lowers the interfacial tension between the two phases (oil and water phases) and serves to prevent the regrouping of the particles formed. The adsorption at the interface between water and air reduces the surface tension [48]. When the concentration of stabilizer is increased, finer emulsion could be formed. Consequently, the size of microcapsules will be reduced and their size distribution will be narrower [44, 49]. However, an increase in nanoparticle formation will also occur, which will cause problem during separation by filtration [23]. Another study has reported the observed pH variation when concentration of EMA was increased [50].

The effect of emulsifier concentrations has been investigated. At low emulsifier concentration (<2 wt. %), agglomeration of particles tends to occur and the microcapsules were reported to have irregular shapes and sizes. Increasing the concentration to 3 wt. % has improved the size distribution. However, at much higher emulsifier levels, the microcapsules might become very small and with uneven size distribution [51]. Overall, it is necessary to optimize the concentration of emulsifier to obtain the desired size, shape, and a good yield.

The successful production of desired microcapsules is affected by the method of production. It is difficult to generalize on the relative importance of the different aspects of the procedure. The relative viscosities of the organic (alkyd/ oils/ others) and the aqueous phases have been reported to be significant parameters [28, 52]. Only the viscosity of the aqueous phase can be more easily controlled or adjusted by the use of emulsifier or viscosity adjuster.

On the other hand, the viscosity of the organic phase is determined by the type of core content used. Commonly encapsulated materials such as inks, fragrances, and pesticides are liquids with low viscosity. According to Ghosh (2009), monomers as self-healing reagents should have low viscosity in order to have the ability to flow into the microcracks, once the microcapsules are ruptured [53]. For this purpose, encapsulated healing agents are mostly liquids, which have low viscosity, such as DCPD, linseed oil, and amine. Although some of the commonly available healing agents could be resins of high molecular weights with medium to high viscosity, they need to be diluted with either reactive or non-reactive diluents, prior to encapsulation. For example, DGEBA resin was diluted with 1-butyl glycidyl ether (BGE) as a reactive diluent with 0.2/1 wt. ratio of BGE/DGEBA [44]. In another example, 40 g of E-51 (bisphenol-A epoxy resin) was diluted with 800 mL of sodium polyacrylate prior to encapsulation to produce self-healing epoxy composites [16].

An undiluted epoxy resin, diglycidyl tetrahydro-o-phthalate (DTHP), with 0.36 Pa·s viscosity, was encapsulated for self-healing epoxy composites [10]. In a subsequent work, the effect of using epoxy resins with different viscosities (but with similar epoxide value) on the healing efficiency was studied. EPON 828, Epoxy 731, and Epoxy 711 resins with viscosity values of 12.5 , 0.85 and 0.53 Pa·s, respectively, were used without any dilution. The lowest viscosity epoxy resin (Epoxy 711) achieved the highest mixing efficiency (83.4%), as compared with Epoxy 731 (79.3%) and EPON 828 (63.7%). The healing efficiency would be favored when the encapsulated epoxy prepolymer has lower viscosity, as the healing agent would be able to flow more easily to fill the gaps in the cracks [54].

A work investigated a number of solvents as diluents for epoxy resins (EPON 828 and EPON 862) and concluded that solvents should ideally have dielectric constant (ε) between 5 and 38. The selection is based on three parameters: dielectric constant (ε), boiling point, and flash point. The solvents chosen for their study included chlorobenzene, phenylacetate (PA), and ethyl phenylacetate (EPA). The combination of 60 mL of epoxy resin-15 pph EPA was concluded as the best combination for solvent-promoted self-healing epoxy material [43].

Using PDMS resin, which has high molecular weight and high viscosity, as healing agent, it was diluted with 30 and 53 wt. % xylenes, respectively, prior to encapsulation. The addition of solvent would lower the relative quantity of healing agent delivered. These microcapsules with solvents were compared with microcapsules containing only PDMS resin as control. The control showed best microcapsule production with good shape, free flowing property, and little debris. On the other hand, microcapsules containing solvent were less uniform in shape, with more debris, and had a tendency to cluster. Solvent also lowered the thermal stability of the microcapsules. However, PDMS with 30% xylene showed the highest healing efficiency at certain loading limit and the efficiency decreased when the content loading exceeded 0.3 mg/cm^3. This observation was explained as follows. The strength of PDMS resin increases with molecular weight during polymerization but the viscosity of the core would also increase, thus solvent would be required to improve the flow. The use of solvents generally degrades capsule quality and thermal stability [55].

In a separate study, Nesterova et al. encapsulated different healing agents for epoxy coating, which include linseed oil, 5-ethylidene-2-norbornene (ENB), DGEBA diluted with BGE, and DCPD. They concluded that the stability of microcapsule is very dependent on the core material. Higher stability was observed in microcapsules formed with more viscous cores. This was possibly due to the higher elasticity of more viscous compounds, which can absorb more stress on the shell material during handling of the capsules [36].

The encapsulation process is uniquely dependent on the core used and the process needs to be optimized. A low-viscosity core is preferred for efficient self-healing application and the core viscosity can be adjusted with suitable diluents. However, the inclusion of diluents may or may not have an impact on the production of microcapsules as well as the healing efficiency.

Finally, the agitation rate also has a great influence on the formation of microcapsules. As the stirring rate is reduced, the core/water interfacial surface is lowered. Thus, the reaction between urea and formaldehyde will occur mostly in the aqueous phase instead of on the interfacial surface. This will result in lots of aggregates, which consequently will reduce the amount of well-formed microcapsules [56]. A high agitation rate can produce smaller core droplets and subsequently smaller microcapsules. However, very high agitation rates would lead to more frequent collisions that cause the deposition of PUF and core content on the stirrer and reactor's wall, leading to lower yield of microcapsules [37]. The increase in collisions of the droplets may also lead to more agglomeration of the microcapsules, which also will reduce the yield [57]. Brown et al., in their study, have established a correlation between average diameter and agitation rate, which is linear in log-log scale [37].

3. Conclusion

The development of self-healing smart polymeric materials has attracted much attention during the last decade. The attractive features of such technologies include prolonging the service life-span of the materials and reducing the cost of repair or replacement of failed component. Currently, the most widely employed technique is by embedding microcapsules that contain a healing agent into the bulk polymer matrix. When cracks develop in the polymer matrix, the curing agent is released from the microcapsules to fill the gaps and subsequently cross-link through suitable reactions to repair the cracks. This microcapsule-based approach can be easily integrated in many polymer systems, such as the epoxy composites and coating, and has a vast potential to be explored and applied commercially. Microencapsulation is a robust technique that can be achieved by optimizing the encapsulation method by *in situ* polymerizing the potential shell material with the healing agent in the core. The two most popular amino resins for encapsulation of the core healing agent are urea-formaldehyde (UF) and melamine-formaldehyde (MF).

The microcapsules were synthesized via *in situ* polymerization to form the shell. The water-immiscible core material was dispersed in an aqueous phase that contains urea or melamine or a combination of both, before the calculated amount of formaldehyde is added. The system is heated for several hours at 40–60°C, preliminary reactions occur in the aqueous phase, producing condensation oligomers that deposit on the surface of the dispersed core particles and continue to polymerize gradually to produce a water-insoluble, cross-linked hard polymer shell. The microcapsules could then be separated by sieving and dosed into the epoxy formulations for composite or coating application.

There are several processing parameters that could affect the yield and the size of the microcapsules. These include core/shell weight ratio, the emulsifier type and concentration, pH, and the stirring speed during microencapsulation.

Acknowledgements

We acknowledge the support of the University of Malaya via Grant PG050-2014A in our related works on using palm oil-based alkyds as curing agents for epoxy matrix [58].

Conflict of interest

The authors declare no conflict of interest.

Author details

Seng Neon Gan[1]* and Nurshafiza Shahabudin[2]

1 Chemistry Department, University of Malaya, Kuala Lumpur, Malaysia

2 Chemistry Division, Centre for Foundation Studies in Science, University of Malaya, Kuala Lumpur, Malaysia

*Address all correspondence to: sngan@um.edu.my

References

[1] Ullah H, Azizli KAM, Man ZB, Ismail MC, Khan MI. The potential of microencapsulated self-healing materials for microcracks recovery in self-healing composite systems: A review. Polymer Reviews. 2016;**56**(3):429-485

[2] Blaiszik BJ, Kramer SLB, Olugebefola SC, Moore JS, Sottos NR, White SR. Self-healing polymers and composites. Annual Review of Materials Research. 2010;**40**:179-211

[3] White SR, Sottos NR, Geubelle PH, Moore JS, Kessler MR, Sriram SR, et al. Autonomic healing of polymer composites. Nature. 2001;**409**:794-797

[4] Jin H, Miller GM, Sottos NR, White SR. Fracture and fatigue response of a self-healing epoxy adhesive. Polymer. 2011;**52**(7):1628-1634

[5] Jin H, Miller GM, Pety SJ, Griffin AS, Stradley DS, Roach D, et al. Fracture behavior of a self-healing, toughened epoxy adhesive. International Journal of Adhesion and Adhesives. 2013;**44**:157-165

[6] Rule JD, Brown EN, Sottos NR, White SR, Moore JS. Wax-protected catalyst microspheres for efficient self-healing materials. Advanced Materials. 2005;**17**(2):205-208

[7] Kamphaus JM, Rule JD, Moore JS, Sottos NR, White SR. A new self-healing epoxy with tungsten (VI) chloride catalyst. Journal of the Royal Society Interface. 2008;**5**(18):95-103

[8] Yadav J, Satyanarayana M, Balanarsaiah E, Raghavendra S. Phosphomolybdic acid supported on silica gel: A mild, efficient and reusable catalyst for the synthesis of 2, 3-unsaturated glycopyranosides by Ferrier rearrangement. Tetrahedron Letters. 2006;**47**(34):6095-6098

[9] Coope TS, Mayer UF, Wass DF, Trask RS, Bond IP. Self-healing of an epoxy resin using scandium (III) triflate as a catalytic curing agent. Advanced Functional Materials. 2011;**21**(24):4624-4631

[10] Yuan YC, Rong MZ, ZhangMQ, Chen J, Yang GC, Li XM. Self-healing polymeric materials usinge poxy/mercaptan as the healant. Macromolecules. 2008;**41**(14):5197-5202

[11] Pang JW, Bond IP. A hollow fibre reinforced polymer composite encompassing self-healing and enhanced damage visibility. Composites Science and Technology. 2005;**65**(11):1791-1799

[12] Xiao DS, Yuan YC, Rong MZ, Zhang MQ. Self-healing epoxy based on cationic chain polymerization. Polymer. 2009;**50**(13):2967-2975

[13] Jin H, Mangun CL, Stradley DS, Moore JS, Sottos NR, White SR. Self-healing thermoset using encapsulated epoxy-amine healing chemistry. Polymer. 2012;**53**:581-587

[14] Li Q, Mishra AK, Kim NH, Kuila T, Lau K-T, Lee JH. Effects of processing conditions of poly (methylmethacrylate) encapsulated liquid curing agent on the properties of self-healing composites. Composites Part B: Engineering. 2013;**49**:6-15

[15] Li Q, Kim NH, Hui D, Lee JH. Effects of dual component microcapsules of resin and curing agent on the self-healing efficiency of epoxy. Composites Part B: Engineering. 2013;**55**:79-85

[16] Yin T, Rong MZ, Zhang MQ, Yang GC. Self-healing epoxy composites—Preparation and effect of the healant consisting of microencapsulated epoxy and latent curing agent.

Composites Science and Technology. 2007;**67**:201-212

[17] Caruso MM, Delafuente DA, Ho V, Sottos NR, Moore JS, White SR. Solvent-promoted self-healing epoxy materials. Macromolecules. 2007;**40**(25):8830-8832

[18] Zako M, Takano N. Intelligent material systems using epoxy particles to repair microcracks and delamination damage in GFRP. Journal of Intelligent Material Systems and Structures. 1999;**10**(10):836-841

[19] Sauvant-Moynot V, Gonzalez S, Kittel J. Self-healing coatings: An alternative route for anticorrosion protection. Progress in Organic Coating. 2008;**63**(3):307-315

[20] Cho SH, Andersson HM, White SR, Sottos NR, Braun PV. Polydimethylsiloxane-based self-healing materials. Advanced Materials. 2006;**18**(8):997-1000

[21] Toohey KS, Sottos NR, Lewis JA, Moore JS, White SR. Self-healing materials with microvascular networks. Nature Materials. 2007;**6**(8):581-585

[22] Suryanarayana C, Rao KC, Kumar D. Preparation and characterization of microcapsules containing linseed oil and its use in self-healing coatings. Progress in Organic Coating. 2008;**63**(1):72-78

[23] Nesterova T, Dam-Johansen K, Pedersen LT, Kiil S. Microcapsule-based self-healing anticorrosive coatings: Capsule size, coating formulation, and exposure testing. Progress in Organic Coating. 2012;**75**(4):309-318

[24] Boura SH, Peikari M, Ashrafi A, Samadzadeh M. Self-healing ability and adhesion strength of capsule embedded coatings—Micro and nano sized capsules containing linseed oil. Progress in Organic Coating. 2012;**75**(4): 292-300

[25] Samadzadeh M, Boura SH, Peikari M, Ashrafi A, Kasiriha M. Tung oil: An autonomous repairing agent for self-healing epoxy coatings. Progress in Organic Coating. 2011;**70**(4):383-387

[26] Jackson AC, Bartelt JA, Marczewski K, Sottos NR, Braun PV. Silica-protected micron and sub-micron capsules and particles for self-healing at the microscale. Macromolecular Rapid Communications. 2011;**32**(1): 82-87

[27] Cho SH, White SR, Braun PV. Self-healing polymer coatings. Advanced Materials. 2009;**21**(6):645-649

[28] Thies C. Microencapsulation. In: Kirk-Othmer Encyclopedia of Chemical Technology. New Jersey, US: Wiley Blackwell, John Wiley & Sons, Inc; 2005

[29] Lamprecht A, Bodmeier R. Microencapsulation. In: Ullman's Encyclopedia of Industrial Chemistry. Weinheim: Wiley VCH Verlag GmbH & Co. KGaA; 2010

[30] Samadzadeh M, Boura SH, Peikaria M, Kasiriha SM, Ashrafic A. A review on self-healing coatings based on micro/ nanocapsules. Progress in Organic Coatings. 2010;**68**:159-164

[31] Ellis B. Chemistry and Technology of Epoxy Resins. Netherlands: Springer-Science+Business Media, B.V; 1993

[32] Ibeh CC. Amino and furan resins. In: Handbook of Thermoset Plastics. Williams Andrew Publishing; 1998

[33] Diem H, Matthias G, Wagner RA. Amino resins. In: Ullmann's Encyclopedia of Industrial Chemistry. Weinheim: Wiley-VCH Verlag GmbH & Co. KGaA; 2010

[34] Singha AS, Thakur VK. Effect of fibre loading on urea-formaldehyde matrix based green composites. Iranian Polymer Journal. 2008;**17**(11):861-873

[35] Murphy EB, Wudl F. The world of smart healable materials. Progress in Polymer Science. 2010;**35**(1-2):223-251

[36] Nesterova T, Dam-Johansen K, Kiil S. Synthesis of durable microcapsules for self-healing anticorrosive coatings: A comparison of selected methods. Progress in Organic Coating. 2011;**70**(4):342-352

[37] Brown EN, Kessler MR, Sottos NR, White SR. In situ poly(urea-formaldehyde) microencapsulation of dicyclopentadiene. Journal of Microencapsulation. 2003;**20**(6):719-730

[38] Liu X, Sheng X, Lee JK, Kessler MR. Synthesis and characterization of melamine-urea-formaldehyde microcapsules containing ENB-based self-healing agents. Macromolecular Materials and Engineering. 2009;**294**(6-7):389-395

[39] Tong X-M, Zhang T, Yang M-Z, Zhang Q. Preparation and characterization of novel melamine modified poly (urea–formaldehyde) self-repairing microcapsules. Colloids and Surfaces A: Physicochemical and Engineering Aspects. 2010;**371**(1): 91-97

[40] Then S, Gan SN, Kasim NHA. Performance of melamine modified urea–formaldehyde microcapsules in a dental host material. Journal of Applied Polymer Science. 2011;**122**(4):2557-2562

[41] Then S, Gan SN, Kasim NHA. Optimization of microencapsulation process for self-healing polymeric material. Sains Malaysiana. 2011;**40**(7):795-802

[42] Noh H, Lee J. Microencapsulation of self-healing agents containing a fluorescent dye. Express Polymer Letters. 2013;**7**:88-94

[43] Blaiszik BJ, Caruso MM, McIlroy DA, Moore JS, White SR, Sottos NR. Microcapsules filled with reactive solutions for self-healing materials. Polymer. 2009;**50**:990-997

[44] Yuan L, Liang G, Xie JQ, Li L, Guo J. Preparation and characterization of poly(urea-formaldehyde) microcapsules filled with epoxy resins. Polymer. 2006;**47**(15):5338-5349

[45] Guo H, Zhao X. Preparation of microcapsules with narrow-size distribution by complex coacervation: Effect of sodium dodecyl sulphate concentration and agitation rate. Journal of Microencapsulation. 2008;**25**(4):221-227

[46] Tiarks F, Landfester K, Antonietti M. Preparation of polymeric nanocapsules by miniemulsion polymerization. Langmuir. 2001;**17**(3):908-918

[47] Song JK, Choi HJ, Chin I. Preparation and properties of electrophoretic microcapsules for electronic paper. Journal of Microencapsulation. 2007;**24**(1): 11-19

[48] Salaün F, Devaux E, Bourbigot S, Rumeaud P. Influence of process parameters on microcapsules loaded with n-hexadecane prepared by in situ polymerization. Chemical Engineering Journal. 2009;**155**:457-465

[49] Zhu DY, Rong MZ, Zhang MQ. Preparation and characterization of multilayered microcapsule-like microreactor for self-healing polymers. Polymer. 2013;**54**(16):4227-4236

[50] Fan C, Zhou X. Influence of operating conditions on the surface morphology of microcapsules prepared by in situ polymerization. Colloids and Surfaces A: Physicochemical and Engineering Aspects. 2010;**363**(1): 49-55

[51] Ting Z, Min Z, Xiao-Mei T, Feng C, Jian-Hui Q. Optimal preparation and characterization of poly (urea–formaldehyde) microcapsules. Journal of Applied Polymer Science. 2010;**115**(4):2162-2169

[52] Jyothi NVN, Prasanna PM, Sakarkar SN, Prabha KS, Ramaiah PS, Srawan G. Microencapsulation techniques, factors influencing encapsulation efficiency. Journal of Microencapsulation. 2010;**27**(3):187-197

[53] Ghosh SK. Self-Healing Materials: Fundamentals, Design Strategies, and Applications. Weinheim: Wiley-VCH Verlag GmbH & Co. KGaA; 2009

[54] Yuan YC, Rong MZ, Zhang MQ, Yang GC. Study of factors related to performance improvement of self-healing epoxy based on dual encapsulated healant. Polymer. 2009;**50**(24):5771-5781

[55] Mangun C, Mader A, Sottos N, White S. Self-healing of a high temperature cured epoxy using poly (dimethylsiloxane) chemistry. Polymer. 2010;**51**(18):4063-4068

[56] Cosco S, Ambrogi V, Musto P, Carfagna C. Properties of poly (urea-formaldheyde) microcapsules containing an epoxy resin. Journal of Applied Polymer Science. 2007;**105**(3):1400-1411

[57] Chen M, Liu J, Liu Y, Guo C, Yang Z, Wu H. Preparation and characterization of alginate–N-2-hydroxypropyl trimethyl ammonium chloride chitosan microcapsules loaded with patchouli oil. RSC Advances. 2015;**5**(19):14522-14530

[58] Shahabudin S, Yahya R, Gan SN, Sonsudin F. Curing of alkyd/epoxy blend in self-healing coating. High Performance Polymers. 2018;**30**(8):1009-1015

Chapter 6

Microencapsulation and Its Uses in Food Science and Technology: A Review

Pedro Henrique Rodrigues do Amaral, Patrícia Lopes Andrade and Leilane Costa de Conto

Abstract

Microencapsulation is a group of technologies aiming to produce small particles called microcapsules that can be released at a specific speed under certain conditions. Microencapsulation technology is used in the pharmaceutical, agrochemical, and food industries; however, microcapsule production is most challenging for applications in the food industry owing to the high costs of the technique, which may make the final product too expensive. Common methods for microencapsulation include spray-drying and coacervation, and different wall materials and filling materials can be used for both techniques. In this review, we summarize current methodologies used for microencapsulation, with a focus on applications in the food industry.

Keywords: microencapsulation, food industry, nutrient enrichment, wall material, core

1. Introduction

Currently, food manufacturers and scientists worldwide are aiming to identify and characterize foods that can be used as sources of beneficial nutrients to promote the health and well-being of consumers. Based on this new paradigm, the development of new food products must combine novel technologies with the use of traditional methods to the control bio-accessibility of certain components in foods. As the interactions among health, nutrition, and genetics are clarified, this approach will become increasingly important.

One effective method for achieving these aims is microencapsulation [1], which was used as early as 1930. The first products containing microencapsulated materials were successfully fabricated in 1954. This advancement promoted further research on the use of microencapsulation in the pharmaceutical industry, wherein researchers found they could use these techniques to achieve controlled release of drugs in the body or in specific organs. Thus, pharmaceutical companies were crucial for developing improved techniques for microencapsulation [2]. In the 1960s, the first studies of microencapsulation in food technology were performed using essential oils; scientists attempted to prevent lipid oxidation, volatile compound losses, and aroma-controlled release. Subsequently, many more studies regarding microencapsulation of food products were published [3].

The goals of microencapsulation are to coat an active compound (core) by an encapsulating agent, also known as wall material, which will isolate the active material, thereby protecting the active material from adverse changes or to hide sensory properties that are not appreciated by consumers. The isolation provided by the encapsulating material will break under the application of a specific stimulus (e.g., pH or heat), releasing the active substance in the specific target location or under ideal conditions [2].

In this review, we summarize the latest applications of microencapsulation and microcapsule production methods in the food industry.

2. Microencapsulation in the food industry

The techniques for producing microcapsules are significantly more challenging in the food industry than in other industries because the sensory qualities of foods cannot be compromised by the addition of encapsulated components. Furthermore, food matrices are more complex than those used in pharmaceutical and cosmetic industries. Moreover, in the food industry, microcapsules must be ingested orally, resist the adverse conditions of the gastrointestinal tract, and exhibit mucoadhesive properties [1]. Several different methods for microcapsule production have been developed, and microcapsules can be fabricated using various materials, which are chosen depending on the function of the microcapsules [4].

Microencapsulation is used to reduce adverse aromas, volatility, and reactivity of food products and to provide food products with greater stability when exposed to adverse conditions (e.g., light, O_2, and pH) [5, 6]. Favaro-Trindade et al. [1] stated that microencapsulation is used in the food industry to reduce the reactivity of the active material in the external environment, reduce the speed of losses and evaporation of the core material into the medium, improve food handling, provide controlled release of the active product, mask unpleasant odor and taste, and allow the encapsulated material to be distributed in a food formulation homogeneously. However, microencapsulation is associated with dramatically increased costs of production, which may limit the economic viability of the method.

Notably, consumers are becoming increasingly aware of the importance of consuming meals that benefit health. Thus, products are being developed to provide health benefits to consumers; microencapsulation of various active compounds, such as vitamins, minerals, essential oils, and omega-3 polyunsaturated fat acids, among others, may be used to protect these compounds from nutrient loss and oxidation reactions and to hide sensory characteristics [2]. Therefore, while there are a wide range of applications of microencapsulation in the food industry, more studies are needed to determine the effectiveness of microencapsulation and the consumer acceptance of products manufactured using microencapsulation [7].

2.1 Microencapsulation processes

Microencapsulation is the science of trapping components (core or active) into a secondary material (encapsulant, wall material, carrier, or cover), producing small solid particles (1–500 μm in size) [8]. These particles are able to release their contents at a specified rate or under specific conditions [1].

The first step in microencapsulation consists of mixing the active material with the encapsulant material, making an emulsion. The mixture can be made with one or two agents. The mixture is then dried, producing microcapsules of different diameters and forms depending on the preparation method and materials used [7].

Physicochemical methods (simple or complex coacervation, separation of organic phase, and liposomal wrapping), physical methods (spray-drying, spray chilling, spray coating, fluidized bed, extrusion, centrifugation with multiple orifices, co-crystallization, and lyophilization), and chemical methods (interfacial polymerization and molecular inclusion) have been developed for microencapsulation [9].

Techniques and materials for microencapsulation are described in **Table 1** [1].

The methods most used by the food industry and which deserve attention are described below.

2.1.1 Coacervation

The oldest microencapsulation technique and one of the most widely used techniques is coacervation, which involves macromolecular aggregates that form a colloidal system with two existing phases: one that is rich in colloids (coacervate) and one that is poor in colloids (supernatant). This method is performed by depositing the encapsulating agent around the active compound through physicochemical changes, such as temperature, polarity, pH, or ionic strength [2, 6].

Methods for encapsulation	Encapsulated materials
Physical methods	
Stationary extrusion	Liquid/solid/gas
Submerged nozzle	Liquid/solid/gas
Centrifugal extrusion	Liquid/solid/gas
Vibrant nozzle	Liquid/solid/gas
Spray-drying	Liquid/solid
Rotating disc	Liquid/solid
Pan coating	Solid
Air suspension	Solid
Spray chilling and spray cooling	Liquid/solid
Fluidized bed	Solid
Co-crystallization	Liquid/solid
Lyophilization	Liquid
Chemical methods	
Interfacial polymerization	Liquid/solid
Molecular inclusion	Liquid
In situ polymerization	Liquid/solid
Physical-chemical methods	
Simple coacervation	Liquid/solid
Complex coacervation	Liquid/solid
Liposomes	Liquid/solid
Lipospheres (solid lipid nanoparticles and nanostructured lipid carriers)	Liquid/solid
Evaporation of the solvent	Liquid/solid

Table 1.
Methods and kind of materials utilized for food products encapsulation.

Coacervation occurs when medium changes make the wall material form polymeric chain units, which then interact with others close chains, forming aggregates. After this step, the aggregates interact with each other through high-intensity attraction forces. Consequently, the aggregated polymer chains will be deposited around the droplets of the hydrophobic phase dispersed in the emulsion, forming a protective film [4].

The microcapsules obtained by coacervation can be classified as mononuclear or multinuclear according to their internal structure. When a drop of core material is encapsulated by coacervation, the particle formed is mononuclear; multinuclear particles are formed by aggregation of various mononuclear microcapsules. Multinuclear microcapsules have a matrix structure, and the core material can be released slowly unless the wall has been broken. However, mononuclear microcapsules have a vessel structure and release all their contents quickly. These particles are also irregular in structure because the wall material is not equally distributed over the surface of the core drop. The thinnest part of the wall layer will be more susceptible to disruption and release of the core.

Therefore, multinuclear microcapsules have greater controlled release and are produced more easily than mononuclear microcapsules [10].

The microcapsules produced by coacervation may have small diameters ranging from 1 to 500 μm for complex coacervation and from 20 to 500 μm for simple coacervation [1]. An example is presented in **Figure 1**. When using lipophilic materials with a hydrophilic coating, high encapsulation efficiency (85–90%) is generally observed [11–15].

Complex coacervation has been used for microencapsulation of sensitive microorganisms and compounds, such as probiotics bacteria, omega-3 products, and bioactive compounds [16–18].

2.1.2 Spray-drying

The use of spray-drying for microencapsulation is another widely used technique due to its low-cost and easy application [19]. Spray-drying technique is used in the food industry for microencapsulating juice, pulp, and vegetal extracts [20, 21], probiotics [22], and fish oil [23].

During spray-drying, a homogeneous mixture of the active material and wall material in aqueous or organic solution is subjected to a hot airstream that promotes the evaporation of the solvent drying the microcapsules. Thus technique generates no solvent residues and does not require a washing process. However, the use of

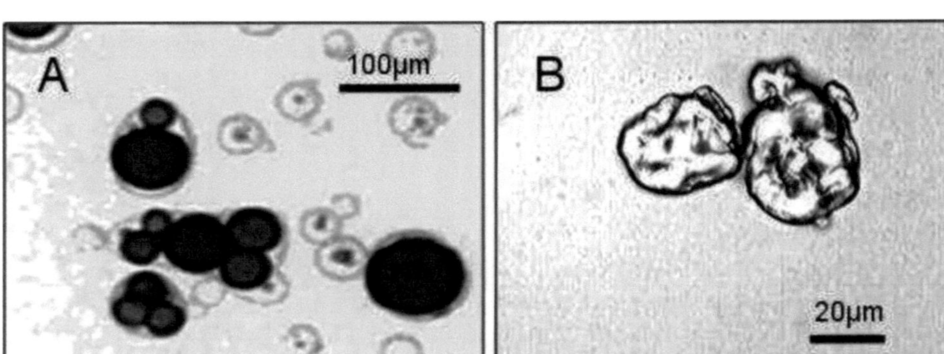

Figure 1.
Microcapsules obtained by complex coacervation with gelatin/gum arabic (A) and soybean protein (B). Source: Author.

high temperatures may compromise the integrity of the core and wall materials. This microcapsule production process has a high efficiency rate, which can be affected by the concentration of the wall material, the system speed, and the feed temperature [24]. Moreover, spray-drying is more widely used than other methods owing to its relatively low cost and capacity for large-scale production [25].

However, according to Kolanowski et al. [11], spray-drying results in porous particles, and this characteristic may increase the susceptibility of the core material to oxidation. Additional disadvantages include the requirement for expensive equipment and the irregularity of the produced microcapsules [6, 24].

Table 2 shows some recent studies about spray-drying application on food microencapsulation.

2.1.3 Fluidized bed

In fluidized bed encapsulation, while the particles of the core material are suspended, the wall material is atomized into the chamber, depositing on the core. When the particles reach the top of the ascending column, they are released into a descending column of air which releases them back into the fluidized bed, where they are again coated, dried, and hardened, ensuring a uniform coating. Fluidized bed encapsulation is one of the few technologies that allow particles to be coated with any wall material (polysaccharides, proteins, emulsifiers, fats, etc.). This method has been used, for example, to isolate iron from ascorbic acid in multivitamin formulations or to encapsulate salt and acidulants avoiding, this way, the interaction of such ingredients with others [34].

Regardless of the method for microencapsulation, release of the core material depends on various factors, including pH, temperature, diffusion, medium solubility, mechanical rupture, and biodegradation. Additionally, the thickness of the encapsulating material may alter the stability and permeability of the microcapsules [1].

2.1.4 Molecular inclusion

One of the most promising possibilities of flavor stabilization is the formation of inclusion complex (molecular encapsulation) with β-cyclodextrin. Szente and

Paper	Source
Flavonoid microparticles by spray-drying: influence of enhancers of the dissolution rate on properties and stability	Sansone et al. [26]
Microencapsulation of linseed oil by spray-drying for functional food application	Gallardo et al. [27]
Optimization of microencapsulation of fish oil with gum arabic/casein/beta-cyclodextrin mixtures by spray-drying	Li et al. [28]
Retention of saffron bioactive components by spray-drying encapsulation using maltodextrin, gum arabic, and gelatin as wall materials	Rajabi et al. [29]
Spray-drying microencapsulation of synergistic antioxidant mushroom extracts and their use as functional food ingredients	Ribeiro et al. [30]
Spray-drying microencapsulation of cinnamon infusions (*Cinnamomum zeylanicum*) with maltodextrin	Santiago-Adame et al. [31]
Influence of different combinations of wall materials on the microencapsulation of jussara pulp (*Euterpe edulis*) by spray-drying	Santana et al. [32]
Sulfur aroma compounds in gum arabic/maltodextrin microparticles	Uekane et al. [33]

Table 2.
Spray-drying studies for microencapsulated food products.

Szejtli [35] investigating the stabilization of natural and synthetic coffee compounds with β-cyclodextrin, and thermal stability of this carbohydrate, observed the molecular encapsulation with natural and synthetic coffee compounds. They also noted that β-cyclodextrin is thermally destroyed at 260°C.

Inclusion compounds of β and γ-cyclodextrins with essential oils of lemon, orange, and camomile have been studied. Lemon and orange oils resulted in the union with β and γ-cyclodextrin. With camomile oil, the complex observed was only with γ-cyclodextrin [36].

2.2 Wall materials

The wall material should be able to form a film that is cohesive with the core material, be chemically compatible and nonreactive with the core material, and provide the desired coating properties, for example, strength, flexibility, impermeability, and stability [37]. In order to be able to be applied in food, the wall material must be food grade, biodegradable, and capable of forming a barrier between the active agent and the medium [19].

Importantly, some core materials are insoluble in aqueous solutions and may not easily form emulsions [38]. Specific proteins may function as emulsifiers, and polysaccharides contribute to the stability of emulsions; the interactions between proteins and carbohydrates can also help stabilize the emulsions.

Among the polysaccharides utilized as wall materials, gum arabic, maltodextrins, and modified starches are the most usual because of the high molecular weight and the high glass transition temperature [19]. However, other polysaccharides are also used, such as carrageenan, carboxymethylcellulose (CMC), and chitosan.

2.2.1 Polysaccharides

Gums are a group of polysaccharides and polysaccharide derivatives obtained from plants or secreted by bacteria and are commonly used for microcapsule production in the food industry.

Gum arabic has low viscosity in water, provides good retention of volatile compounds (>85%), and protects the core material from oxidation, which is crucial for microencapsulating essential oils and volatile substances [7]. Gum arabic has advantages for having this property emulsifier in a wide pH range, as well as other texturing, training film around the droplets and binding properties [38]. Conto et al. [17] studied the complex coacervation of soy proteins with gum arabic (GA); Renard et al. [39] worked with vitamin E microencapsulated on β-lactoglobulin/GA matrix.

Alternatively, alginate can be used for microencapsulation. This material forms strong, elastic gels with a distinct three-dimensional network. The gel network and homogeneity depends on the cation concentration; excess Ca^{2+} may result in multiple alginate chains having different physicochemical properties.

Alginate can also be used to produce microcapsules and cell immobilization through ionotropic gelation, which involves dropping the concentrated alginate solution into calcium chloride solution, externally gelling the polymer into a microcapsule. The size of the microcapsules formed using external gelation is governed by the size of droplets formed during the extrusion process [40] and ranges from tens of microns to millimeter size. Less commonly, microcapsules may be formed by internal gelation, in which the alginate in solution contains calcium carbonate [41].

Another common use of alginate microcapsules is to reduce the viability losses of probiotic bacteria, like *Bifidobacterium* and *Lactobacillus*. Some works with probiotics alginate encapsulation are presented by Cook et al. [42] and are summarized in **Table 3**.

Encapsulation material	Bacteria	Reference apud Cook et al. [42]
Alginate	*Lactobacillus acidophilus*	Chandramouli et al. [40]
Alginate coated with palm oil and poly-L-lysine	8 different *Lactobacilli* and *Bifidobacteria*	Ding et al. [43]
Alginate and xanthan gum	*Lactobacillus acidophilus*	Kim et al. [44]
Alginate coated with either chitosan, alginate, or poly-L-lysine-alginate	*Lactobacillus acidophilus*, *Bifidobacterium bifidum*, *Lactobacillus casei*	Krasaekoopt et al. [45]
Alginate	*Lactobacillus casei*	Mandal et al. [46]
Alginate Alginate and pectin	*Lactobacillus casei*	Sandoval-Castilla et al. [41]
Alginate coated with whey protein	*Lactobacillus plantarum*	Gbassi et al. [47]
Alginate coated with chitosan, Sureteric, or Acryl-EZE	*Bifidobacterium animalis*	Liserre et al. [48]
Alginate coated with chitosan	*Bifidobacterium breve*	Cook et al. [42]

Table 3.
Overview of literature available on the alginate encapsulation of probiotics cited by Cook et al. [42].

Carrageenans are widely used as thickening and stabilizing agents. Previous studies have reported microcapsules produced by carrageenan and oligochitosan polymer, but most reports have described the use of carrageenan for the encapsulation of microbial cells due to its capacity for gelation with the change in temperature from 40 to 45°C [49], suggesting the potential for use in probiotic foods.

Starch and modified starch can also be used as a wall material owing to its low viscosity, outstanding retention volatility (>93%), and ability to stabilize the emulsion with the core material [7]. Starches and their derivatives have been applied for the microencapsulation of vitamins, such as ascorbic acid [50, 51]. Maltodrextrin, which is inexpensive and has low hygroscopicity, can be used to prevent particle agglomeration [17] and has antioxidant effects [7].

Chitosan is also commonly used as a gelation agent in the food and pharmaceutical industries. Chitosan also allows concurrent cell permeabilization and immobilization; thus, chitosan-containing complexes of coacervated capsules have been widely explored [52].

Additionally, carboxymethylcellulose (CMC), an anionic water-soluble polymer, is used as an industrial agent owing to its capacity as a thickener, suspending agent, stabilizer, and binder. CMC forms resistant films that can protect against organic solvents, oils, and greases [53].

2.2.2 Protein films

Protein films are excellent oxygen and aroma barriers and can be used to produce microcapsules using coacervation techniques [54] or double emulsions with subsequent reticulation using glutaraldehyde or heat gelation [55]. Usually, proteins have been utilized with other biopolymers; some examples are presented in **Table 4**.

Whey proteins (WPC and WPI) have also been investigated as wall materials for microencapsulation. For example, whey protein has been used for encapsulation of volatile and nonvolatile materials [56], typically through spray-drying, complex coacervation, heat gelation, and enzymatic gelation [57, 58]. Combinations such as whey protein isolate/gum arabic [59], β-lactoglobulin/pectin [60], β-lactoglobulin (b-Lg)/κ-carrageenan [61], whey protein/chitosan/gum arabic [62], and milk protein products/xanthan [63] are frequently used.

Despite these studies, proteins from plant sources have not been commonly used as carrier or wall materials in microencapsulation applications owing to limitations related to heat instability and organic solvent sensitivity. However, the use of reticulating agents to convert the proteins into a more stable form may improve their industrial applicability [64].

Additionally, soy proteins have the benefits of renewability, low-cost, and healthful effects [65]. Soy protein has high compatibility with gum arabic. SPI has been successfully used for microencapsulation of casein hydrolysate by spray-drying [66], orange essential oil by complex coacervation [13], and fish oil by enzymatic gelation [57]. **Figure 2** presents microcapsules obtained with SPI by complex coacervation and enzymatic gelation.

Gelatin can also be used as a foaming agent, emulsifier, and humectants in food, pharmaceutical, medical, and technical application sowing to its surface-active properties. Type A gelatin has a high isoelectric point and can form oil/water emulsions with positive charges at a wide range of pH values [67].

2.3 Core materials

Among core materials, essential oils are highly unstable and are sensitive to variations in light, air, temperature, and humidity. Therefore, new methods are needed to protect oils against these changes in order to increase their shelf life and their chemical stability under adverse conditions [1, 7].

Wall material	Core material	Source
Gelatin/gum arabic	Soybean oil, olive oil, and peanut oil	Rabišková, Valasková [68]
Gelatin/gum arabic	EPA	Lamprecht, Schäfer, Lehr [69]
Gelatin/gum arabic	Fish oil	Jouzel et al. [70]
Whey protein/gum arabic	Sunflower oil, lemon, and orange essential oil	Weinbreck, Minor, DeKuif [58]
Hydroxpropyl methylcellulose	Fish oil	Wu, Chai, Chen [71]
Gelatin/gum arabic	Oleoresin and soybean oil	Alvim [72]
Gelatin/gum arabic	Baking flavor	Yeo et al. [73]
Gelatin/pectin/gum arabic	Oils	Prata [74]
Gelatin/gum arabic	Peppermint oil	Dong et al. [10]
Gelatin	Stigmasterol	Oliveira [75]
SPI/pectin	Casein hydrolyzate	Medanha et al. [76]
b-Lg/pectin	DHA	Zimet, Livney [77]
Gelatin/polyphosphate	Fish oil ethyl ester	Barrow, Nolan, Holub [78]
Gelatin/gum arabic	Flavors	Leclercq, Milo, Reineccius [79]
Gelatin/gum arabic	Soybean oil and paprika resin oil	Célis [80]
Gelatin/gum arabic	1-Dodecanol (C12OH)	Kong et al. [81]
HPMC/NaCMC/SDS	Sunflower oil	Katona, Sovilj, Petrovic [82]
SPI/gum arabic	Orange essential oil	Jun-Xia, Hai-Yan, Jian [13]

Table 4.
Overview of literature available on the proteins as encapsulant material.

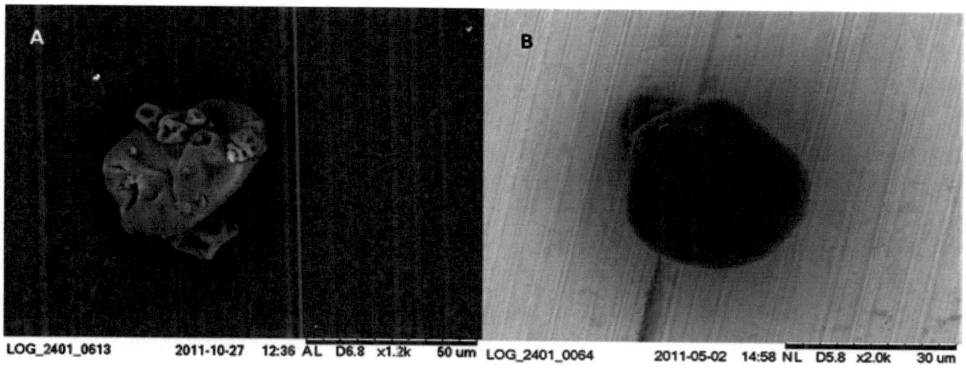

Figure 2.
MEV of SPI microcapsules obtained by complex coacervation (A) and enzymatic gelation (B). Source: Author.

Vitamins and minerals are generally added to foods to increase nutritional value, such as cereals, dairy products, infant foods, etc. However, these compounds can cause the food to taste unpleasant or may react with other food constituents, changing their sensory characteristics. Therefore, microencapsulation is widely used to protect vitamins and minerals against adverse conditions, such as temperature and humidity, to prevent undesirable reactions in food [1].

Microorganism microencapsulation has been applied to allow reuse of bacteria during the production of lactic acid and fermented milk products; increase production and cell concentrations in reactors; provide protection against oxygen gas, freezing temperatures, and the unfavorable pH of gastric juices and other acids; remove cell sand stop acidification; provide greater stability and maintain the viability of cultures during product storage; and increase their useful life [1].

Microencapsulation is widely used for enzyme immobilization, allowing reuse of enzymes and providing enzymes with superior stability; these features also reduced costs associated with the relevant processes. Moreover, microencapsulation for immobilization of enzymes is simple and permits the production of microcapsules having a variety of compositions [83].

Many studies have shown that consumption of omega-3 polyunsaturated fatty acids provides multiple health benefits, including reducing the risk of cardiovascular disease. Polyunsaturated fatty acids of the omega-3 group are mainly found in marine animals, such as plankton and fish in cold and deep waters, and fish oil has been the traditional source of these fatty acids. However, fish oil has an undesirable flavor. Thus, microencapsulation has been used for incorporation of fish oil as the core material, hiding the unpleasant sensory characteristics of the oil [17].

In studies with omega-3 microcapsules applied in food products, Chavez-Servín et al. [84] examined the addition of microencapsulated omega-3 fatty acids in infant formulas. Lysine and lactose degradation were observed; however, it was determined that the microcapsules did not affect the sensory acceptance of the final product. Moreover, Yep et al. [85] applied omega-3 microcapsules in bakery products and evaluated the effects of consumption of small daily doses of omega-3 fatty acids by intake of commercial bread compared to supplementation with capsules. They concluded that the effects depended on the amount of EPA and DHA in the blood of the individuals studied. Serna-Saldivar et al. [86] determined the shelf life of bread enriched with DHA and microencapsulated fish oil, showing that the development of off-flavors occurred more quickly in the breads containing liquid fish oil. Davidov-Pardo et al. [87] also observed changes in technological and sensory characteristics of breads containing omega-3 microcapsules.

The encapsulation of acids such as ascorbic, citric, fumaric, and lactic acids is usually carried out to avoid oxidation and allow them to be dissolved under specific conditions. Three specific applications stand out in this case: the dough improver, because the encapsulated ascorbic acid is often used to alter dough strength and improve slicing properties, color, and texture of baked products, being released during the proofing and baking stages, the aroma complement, and as an auxiliary agent in the meat processing, allowing the desired cured meat pigments to form [88].

Some natural colorants, such as *urucum*, β-carotene, and *Curcuma*, have solubility problems. It can be solved by encapsulation processes, which make them easier to handle during the process and improve the solubility and oxidation stability. Another advantage that can be associated with its use is in the shelf life extension, which can exceed 2 years, compared to the 6 months for non-encapsulated ones [88].

3. Conclusion

Foods and other substances microencapsulated exhibit wide applicability, being an effective and extremely important tool in the preservation of various nutritional components, microorganisms, enzymes, dyes, etc., protecting food and other products from the most aggressive processing methods.

Several materials can be used as encapsulants, the most common being carbohydrates and some proteins, due to their higher affinity with various types of materials to be encapsulated. There are several methods of encapsulation by physical, chemical, and physicochemical, the most used being atomization, fluidized bed, and coacervation.

Despite the wide applicability, encapsulation has found little space in the food industry because of the cost. While the pharmaceutical and cosmetic sectors often support the use of high-cost techniques, the food industry works with lower profit margins, reducing production costs. In addition, industries often have strong resistance to the adoption of new technologies, due to the cost of implementation and the need for training.

Development of methodologies for incorporation of functional compounds in foods is needed to improve the health benefits and marketability of foods. Finally, microencapsulation of nutrients is a relatively new technology in the food industry, and further studies are needed to determine how to apply this technology most effectively.

Acknowledgements

This work was partly funded by the Federal Institute of Triangulo Mineiro—Campus Uberlândia and a FAPEMIG research scholarship.

Author details

Pedro Henrique Rodrigues do Amaral[1], Patrícia Lopes Andrade[1] and Leilane Costa de Conto[2]*

1 Triangulo Mineiro Federal Institute, Uberlândia, Brazil

2 Federal Institute of Santa Catarina, Urupema, Brazil

*Address all correspondence to: leilanedeconto@gmail.com

References

[1] Favaro-Trindade CS, De Pinho SC, Rocha GA. Revisão: Microencapsulação de ingredientes alimentícios. Brazilian Journal of Food Technology. 2008;**11**:103-112

[2] Suave J, Dall'agnol EC, Pezzin APT, Silva DAK, Meier MM, Soldi V. Microencapsulação: Inovação em diferentes áreas. Revista Saúde e Ambiente. 2006;7:12-20. ISSN 2175-1641

[3] Gouin S. Microencapsulation: Industrial appraisal of existing technologies and trends. Food Science and Technology. 2004;**15**:330-347. DOI: 10.1016/j.tifs.2003.10.005

[4] Leimann FV. Microencapsulação de óleo essencial de capim-limão utilizando o processo de coacervação simples [thesis]. Florianópolis: UFSC; 2008

[5] Müller PS. Microencapsulação do óleo essencial de laranja [thesis]. Curitiba: Universidade Federal do Paraná; 2011

[6] Foglio MA, Servat L, Spindola HM, Rodrigues RAF. Microencapsulação: Uma alternativa promissora para preservação de produtos naturais. Revista Fitos Eletrônica. 2013;**5**:52-57. ISSN: 2446-4927

[7] Aburto LC, Tavares D d Q, Martucci ET. Microencapsulação de óleo essencial de laranja. Ciência e Tecnologia de Alimentos. 1998;**18**:45-48. DOI: 10.1590/S0101-20611998000100010

[8] Sanguansri L, Augustin MA. Microencapsulation and delivery of omega-3 fatty acids. Functional food ingredients and nutraceuticals. In: Shi J, editor. Functional Food Ingredients and Nutraceuticals. Boca Raton: CRC Press; 2007. pp. 297-327

[9] Jyothi NVN, Prasanna M, Prabha S, Seetha Ramaiah P, Srawan G, Sakarkar SN. Microencapsulation techniques, factors influencing encapsulation efficiency: A review. The Internet Journal of Nanotechnology. 2009;**3**:1-61. DOI: 10.5580/27bb

[10] Dong ZJ, Toure A, Jia CS, Zhang XM, Xu SY. Effect of processing parameters on the formation of spherical multinuclear microcapsules encapsulating peppermint oil by coacervation. Journal of Microencapsulation. 2007;**24**:634-646. DOI: 10.1080/02652040701500632

[11] Kolanowski W, Jaworska D, Weiâbrodt J, Kunz B. Sensory assessment of microencapsulated fish oil powder. Journal of the American Oil Chemists' Society. 2007;**84**:37-45. DOI: 10.1007/s11746-006-1000-x

[12] Martins IM, Barreiro MF, Coelho M, Rodrigues AE. Microencapsulation of essential oils with biodegradable polymeric carriers for cosmetic applications. Chemical Engineering Journal. 2014;**245**:191-200. DOI: 10.1016/j.cej.2014.02.024

[13] Jun-Xia X, Hai-Yan Y, Jian Y. Microencapsulation of sweet orange oil by complex coacervation with soybean protein isolate/gum arabic. Food Chemistry. 2011;**125**:1267-1272. DOI: 10.1016/j.foodchem.2010.10.063

[14] Comunian TA, Thomazini M, Alves AJG, Matos Junior FE, Carvalho Balieiro JC, Favaro-Trindade CS. Microencapsulation of ascorbic acid by complex coacervation: Protection and controlled release. Food Research International. 2013;**52**:373-379. DOI: 10.1016/j.foodres.2013.03.028

[15] Lima JR, Locatelli GO, Finkler L, Luna-Finkler CL. Incorporação de

Lactobacillus casei microencapsulado em queijo tipo coalho. Ciência & Saúde. 2014;**7**:27-34. DOI: 10.15448/1983-652X.2014.1.15639

[16] Simeoni CP, Etchepare MA, Menezes CR, Fries LM, Menezes MFC, Stefanello FS. Microencapsulation of probiotics: Technological innovation in the food industry. Revista do Centro de Ciências Naturais e Exatas. 2014;**18**:66-75. DOI: 10.5902/2236117013020

[17] Conto LC, Grosso CRG, Gonçalves LAG. Chemometry as applied to the production of omega-3 microcapsules by complex coacervation with soy protein isolate and gum arabic. LWT—Food Science and Technology. 2013;**53**:218-224

[18] Silva TM, Rodrigues LZ, Nunes GL, Codevilla CF, Silva CB, Menezes CR. Encapsulation of bioactive compounds by complex coacervation. Ciência e Natura. 2015;**37**:56-64. DOI: 10.5902/2179-460X19715

[19] Pereira KC, Ferreira DCM, Alvarenga GF, Pereira MSS, Barcelos MCS, Costa JMG. Microencapsulation and release controlled by the diffusion of food ingredients produced by spray drying: A review. Brazilian Journal of Food Technology. 2018;**211**:e2017083. DOI: 10.1590/1981-6723.08317

[20] Da Silva MV, Junior BD, Visentainer JV. Production and characterization of maltodextrins and its application in microencapsulation of food compounds by spray drying. Revista Ciências Exatas e Naturais. 2014;**16**:11-126. DOI: 10.5935/RECEN.2014.01.07

[21] Santana AA, Oliveira RA, Kurozawa LE, Park KJ. Microencapsulação de polpa de pequi por spray drying: Uso de dextrina como agente encapsulante. Engenharia Agrícola. 2014;**34**:980-991. DOI: 10.1590/S0100-69162014000500017

[22] Menezes CR, Barin JS, Chicoski AJ, Zepka LQ, Jacob-Lopes E, Friesi LLM, et al. Microencapsulation of probiotics: Progress and prospects. Ciência Rural. 2013;**43**:1309-1316. ISSN 0103-8478

[23] Kuhn F, Dal Magro J, Raja S, Ferrando M. Encapsulation of fish oil for food use by membrane emulsification followed by atomization. Revista Brasileira de Pesquisa em Alimentos. 2014;**5**:1-8. DOI: 10.14685/rebrapa.v5i1.133

[24] Silva C, Ribeiro A, Ferreia D, Veiga F. Oral delivery system for peptides and proteins: II. Application of microencapsulation methods. Brazilian Journal of Pharmaceutical Sciences. 2003;**39**:1-20. DOI: 10.1590/S1516-93322003000100002

[25] Lee SJ, Ying DY. Encapsulation of fish oils. In: Garti N, editor. Delivery and Controlled Release of Bioactives Food Nutraceuticals. Boca Raton: CRC; 2008

[26] Sansone F, Picerno P, Mencherini T, Villecco F, D'Ursi AM, Aquino RP, et al. Flavonoid microparticles by spray-drying: Influences of enhancers of the dissolution rate on properties and stability. Journal of Food Engineering. 2011;**103**:188-196. DOI: 10.1016/j.jfoodeng.2010.10.015

[27] Gallardo G, Guida L, Martinez V, López MC, Bernhardt D, Blasco R, et al. Microencapsulation of linseed oil by spray drying for functional food application. Food Research International. 2013;**52**:473-482. DOI: 10.1016/j.foodres.2013.01.020

[28] Li J, Xiong S, Wang F, Regenstein JM, Liu R. Optimization of microencapsulation of fish oil with gum arabic, casein, beta-cyclodextrin mixtures by spray drying. Journal of Food Science. 2015;**80**:1445-1452. DOI: 10.1111/1750-3841.12928

[29] Rajabi H, Ghorbani M, Jafari SM, Sadeghi Mahoonak A, Rajabzadeh G. Retention of saffron bioactive components by spray drying encapsulation using maltodextrin, gum arabic and gelatin as wall materials. Food Hydrocolloids. 2015;**51**:327-337. DOI: 10.1016/j.foodhyd.2015.05.033

[30] Ribeiro A, Ruphuy G, Lopes JC, Dias MM, Barros L, Barreiro F, et al. Spray-drying microencapsulation of synergistic antioxidant mushroom extracts and their use as functional food ingredients. Food Chemistry. 2015;**188**:612-618. DOI: 10.1016/j. foodchem.2015.05.061

[31] Santiago-Adame R, Medina-Torres L, Gallegos-Infante JA, Calderas F, Gonzáles-Laredo RF, Rocha-Guzmán NE, et al. Spray drying microencapsulation of cinnamon infusions (*Cinnamomum zeylanicum*) with maltodextrin. Food Science and Technology. 2015;**64**:571-577. DOI: 10.1016/j.lwt.2015.06.020

[32] Santana AA, Cano-Higuita DM, Oliveira RA, Telis VRN. Influence of different combinations of wall materials on the microencapsulation of jussara pulps (*Euterpe edulis*) by spray drying. Food Chemistry. 2016;**212**:1-9. DOI: 10.1016/j.foodchem.2016.05.148

[33] Uekane TM, Costa ACP, Pierucci APTR, Rocha-Leão MHM, Rezende CM. Sulfur aroma compounds in gum arabic/maltodextrin microparticles. Lebensmittel-Wissenschaft und Technologie. 2016;**70**:342-348. DOI: 10.1016/j. lwt.2016.03.003

[34] Azeredo HMC. Encapsulation: Applications to food technology. Alimentos e Nutrição. 2005;**16**:89-97. ISSN 0103-4235

[35] Szente L, Szejtli J. Molecular encapsulation of natural and synthetic coffee flavor with β-cyclodextrin. Journal of Food Science. 1986;**51**:1024-1027

[36] Thoss M, Schawabe L, Fromming KH. Cyclodextrin inclusion compounds of lemon oil, orange oil, hop oil and chamomile oil. PZ Wissenschaft. 1993;**138**:144-148

[37] Bansode SS, Banarjee SK, Gaikwad DD, Jadhav SL, Thorat RM. Microencapsulation: A review. International Journal of Pharmaceutical Sciences Review and Research. 2010;**1**:38-43. ISSN: 0975-6299

[38] Gharsallaoui A, Roudaut G, Chambin O, Voilley A, Saurel R. Applications of spray-drying in microencapsulation of food ingredients: An overview. Food Research International. 2007;**40**:1107-1121. DOI: 10.1016/j.foodres.2007.07.004

[39] Renard D, Robert P, Lavenant L, Melcin D, Popineau Y, Gueguen J, et al. Biopolymeric colloidal carriers for encapsulation or controlled release applications. International Journal of Pharmaceutics. 2002;**242**:163-166. DOI: 10.1016/S0378-5173(02)00143-6

[40] Chandramouli V, Kailasapathy K, Peiris P, Jones M. An improved method of microencapsulation and its evaluation to protect *Lactobacillus* spp. in simulated astric conditions. Journal of Microbiological Methods. 2004;**56**:27-35. DOI: 10.1016/j.mimet.2003.09.002

[41] Sandoval-Castilla O, Lobato-Calleros C, García-Galindo HS, Alvarez-Ramírez J, Vernon-Carter EJ. Textural properties of alginate-pectin beads and survivability of entrapped *Lb. casei* in simulated gastrointestinal conditions and in yoghurt. Food Research International. 2010;**43**:111-117. DOI: 10.1016/j.foodres.2009.09.010

[42] Cook MT, Tzortzis G, Charalampopoulos D, Khutoryanskiy

VV. Microencapsulation of probiotics for gastrointestinal delivery. Journal of Controlled Release. 2012;**162**:56-67. DOI: 10.1016/j.jconrel.2012.06.003

[43] Ding WK, Shah NP. An improved method of microencapsulation of probiotic bacteria for their stability in acidic and bile conditions during storage. Journal of Food Science. 2009;**74**:M53-M61. DOI: 10.1111/j.1750-3841.2008.01030.x

[44] Kim S-J, Cho SY, Kim SH, Song O-J, Shin IIS, Cha DS, et al. Effect of microencapsulation on viability and other characteristics in Lactobacillus acidophilus ATCC 43121. LWT—Food Science and Technology. 2008;**41**:493-500. DOI: 10.1016/j.lwt.2007.03.025

[45] Krasaekoopt W, Bhandari B, Deeth H. The influence of coating materials on some properties of alginate beads and survivability of microencapsulated probiotic bacteria. International Dairy Journal. 2004;**14**:737-743. DOI: 10.1016/j.idairyj.2004.01.004

[46] Mandal S, Puniya AK, Singh K. Effect of alginate concentrations on survival of microencapsulated *Lactobacillus casei* NCDC-298. International Dairy Journal. 2006;**16**:1190-1195. DOI: 10.1016/j.idairyj.2005.10.005

[47] Gbassi GK, Vandamme T, Ennahar S, Marchioni E. Microencapsulation of *Lactobacillus plantarum* spp in an alginate matrix coated with whey proteins. International Journal of Food Microbiology. 2009;**129**:103-105. DOI: 10.1016/j.ijfoodmicro.2008.11.012

[48] Liserre AM, Re MI, Franco B. Microencapsulation of *Bifidobacterium* animalis subsp lactis in modified alginate-chitosan beads and evaluation of survival in simulated gastrointestinal conditions. Food Biotechnology. 2007;**21**:1-16. DOI: 10.1080/08905430701191064

[49] Bartkowiak A, Hunkeler D. Carrageenan-oligochitosan microcapsules: Optimization of the formation process. Colloids and Surfaces B: Biointerfaces. 2001;**21**:285-298. DOI: 10.1016/S0927-7765(00)00211-3

[50] Trindade MA, Grosso CRF. The stability of ascorbic acid microencapsulated in granules of rice starch and in gum arabic. Journal of Microencapsulation. 2000;**17**:169-176. DOI: 10.1080/026520400288409

[51] Uddin MS, Hawlader MNA, Zhu HJ. Microencapsulation of ascorbic acid: Effect of process variables on product characteristics. Journal of Microencapsulation. 2001;**18**:199-209. DOI: 10.1080/02652040010000352

[52] Pegg RB, Shahidi F. Encapsulation, stabilization, and controlled release of food ingredients and bioactives. In: Rahman MS, editor. Handbook of Food Preservation. Boca Raton: CRC Press; 2007. pp. 509-568

[53] Nisperos-Carriedo MO. Edible coatings and films based on polysaccharides. In: Krochta JM, Baldwin EA, Nisperos-Carriedo MO, editors. Edible Coatings and Films to Improve Food Quality. Lancaster: Technomic Publishing Co. Inc.; 1994. p. 305

[54] Conto LC, Fernandes GD, Grosso CRF, Eberlin MN, Gonçalves LAG. Evaluation of the fatty matter contained in microcapsules obtained by double emulsification and subsequent enzymatic gelation method. Food Research International. 2013;**54**:432-438. DOI: 10.1016/j.foodres.2013.07.013

[55] Lee SJ, Rosenberg M. Whey protein-based microcapsules prepared by double emulsification and heat gelation.

LWT—Food Science and Technology. 2000;**33**:80-88. DOI: 10.1006/fstl.1999.0619

[56] Rosenberg M. Milk Derived Whey Protein-based Microencapsulating Agents and A Method of Use. US Patent Number 5601760; 1997

[57] Cho YH, Shim HK, Park J. Encapsulation of fish oil by an enzymatic gelation process using transglutaminase cross-linked protein. Journal of Food Science. 2003;**68**: 2717-2723. DOI: 10.1111/j.1365-2621.2003.tb05794

[58] Weinbreck F, Minor M, De Kruif CG. Microencapsulation of oils using whey protein/gum arabic coacervates. Journal of Microencapsulation. 2004;**21**:667-679. DOI: 10.1080/02652040400008499

[59] Klein M, Aserin A, Svitov I, Garti N. Enhanced stabilization of cloudy emulsions with gum arabic and whey protein isolate. Colloids and Surfaces B: Biointerfaces. 2010;**77**:75-81. DOI: 10.1016/j.colsurfb.2010.01.008

[60] Guzey D, Kim HJ, McClements DJ. Factors influencing the production of o/w emulsions stabilized by β-lactoglobulin-pectin membranes. Food Hydrocolloids. 2004;**18**:967-975. DOI: 10.1016/j.foodhyd.2004.04.001

[61] Gu YS, Regnier L, McClements DJ. Influence of environmental stress on stability of oil-in-water emulsions containing droplets stabilized by β-lactoglobulin-κ-carrageenan membranes. Journal of Colloid and Interface Science. 2005;**286**:551-558. DOI: 10.1016/j.jcis.2005.01.051

[62] Moschakis T, Murray BS, Biliaderis CG. Modifications in stability and structure of whey protein-coated o/w emulsions by interacting chitosan and gum arabic mixed dispersions. Food Hydrocolloids. 2010;**24**:8-17. DOI: 10.1016/j.foodhyd.2009.07.001

[63] Hemar Y, Tamehana M, Munto PA, Singh H. Viscosity, microstructure and phase behaviour of aqueous mixtures of commercial milk protein products and xanthan gum. Food Hydrocolloids. 2001;**15**:565-574. DOI: 10.1016/S0268-005X(01)00077-7

[64] Nesterenko A, Alric I, Silvestre F, Durrieu V. Vegetable proteins in microencapsulation: A review of recent interventions and their effectiveness. Industrial Crops and Products. 2013;**42**:469-479. DOI: 10.1016/j.indcrop.2012.06.035

[65] Tang C-H, Li X-R. Microencapsulation properties of soy protein isolate and storage stability of the correspondingly spray-dried emulsions. Food Research International. 2012;**52**:419-428. DOI: 10.1016/j.foodres.2012.09.010

[66] Molina-Ortiz SE, Mauri A, Monterrey-Quintero ES, Trindade MA, Santana AS, Favaro-Trindade CS. Production and properties of casein hydrolysate microencapsulated by spray drying with soybean protein isolate. LWT—Food Science and Technology. 2009;**42**:919-923. DOI: 10.1016/j.lwt.2008.12.004

[67] Dickinson E, Lopez G. Comparison of the emulsifying properties of fish gelatin and commercial fish proteins. Journal of Food Science. 2001;**66**:118-123. DOI: 10.1111/j.1365-2621.2001.tb15592.x

[68] Rabišková M, Valášková J. The influence of HLB on the encapsulation of oils by complex coacervation. Journal of Microencapsulation. 1998;**15**:747-751. DOI: 10.3109/02652049809008257

[69] Lamprecht A, Schafer UF, Lehr CM. Influences of process parameters

on preparation of microparticle used as a carrier system for O-3 unsaturated fatty acid ethyl esters used in supplementary nutrition. Journal of Microencapsulation. 2001;**18**:347-357. DOI: 10.1080/02652040010000433

[70] Jouzel B, Pennarun A-L, Prost C, Renard D, Poncelet D, Demaimay M. Encapsulation of a lipid precursor, the eicosapentaenoic acid, to study the development of the *Crassostrea gigas* oyster flavours. Journal of Microencapsulation. 2003;**20**(1):35-46. DOI: 10.3109/02652040309178047

[71] Wu KG, Chai XH, Chen Y. Microencapsulation of fish oil by simple coacervation of hydroxypropyl methylcellulose. Chinese Journal of Chemistry. 2005;**23**:1269-1572. DOI: 10.1002/cjoc.200591569

[72] Alvim ID. Produção e caracterização de micropartículas obtidas por spray drying e coacervação complexa e seu uso para alimentação de larvas de peixes [thesis]. Faculdade de Engenharia de Alimentos: Universidade Estadual de Campinas; 2005

[73] Yeo Y, Bellas E, Firestona W, Languer R, Kohane DS. Complex coacervates for thermally sensitive controlled release of flavour compounds. Journal of Agricultural and Food Chemistry. 2005;**53**:7518-7525. DOI: 10.1021/jf0507947

[74] Prata AS. Estudos dos parâmetros físico-químicos envolvidos na formação de microcápsulas produzidas por coacervação complexa [thesis]. Faculdade de Engenharia de Alimentos: Universidade Estadual de Campinas; 2006

[75] Oliveira AB. Microencapsulamento de Estigmasterol proveniente de *Musa paradisiaca* L., Musaceae [thesis]. Ciências Farmacêuticas, Setor de Ciências da Saúde: Universidade Federal do Paraná; 2007

[76] Medanha DV, Ortiz SEM, Favaro-Trindade CS, Mauri A, Monterrey-Quintero ES, Thomazini M. Microencapsulation of casein hydrolysate by complex coacervation with SPI/pectin. Food Research International. 2009;**42**:1099-1104. DOI: 10.1016/j.foodres.2009.05.007

[77] Zimet P, Livney Y. Beta-lactoglobulin and its nanocomplexes with pectin as vehicles for w-3 polyunsaturated fatty acids. Food Hydrocolloids. 2009;**23**:1120-1126. DOI: 10.1016/j.foodhyd.2008.10.008

[78] Barrow CJ, Nolan C, Holub BJ. Bioequivalence of encapsulated and microencapsulated fish-oil supplementation. Journal of Functional Foods. 2009;**1**:38-43. DOI: 10.1016/j.jff.2008.09.006

[79] Leclercq S, Milo C, Reineccius GA. Effects of cross-linking, capsule wall thickness, and compound hydrophobicity on aroma release from complex coacervate microcapsules. Journal of Agricultural and Food Chemistry. 2009;**57**:1426-1432. DOI: 10.1021/jf802472q

[80] Célis FT. Efeito da reticulação induzida pela transglutaminase e o glutaraldeido sobre as propriedades das micropartículas obtidas por coacervação complexa [thesis]. Faculdade de Engenharia de Alimentos: Universidade Estadual de Campinas; 2009. 148 p

[81] Kong XZ, Gu X, Zhu X, Zhang Z. Spreadable dispersion of insect sex pheromone capsules, preparation via complex coacervation and release control of the encapsulated pheromone component molecule. Biomedical Microdevices. 2009;**11**:275-285. DOI: 10.1007/s10544-008-9234-z

[82] Katona JM, Sovilj VJ, Petrovic LB. Microencapsulation of oil by polymer mixture-ionic surfactant interaction

induced coacervation. Carbohydrate Polymers. 2010;**79**:563-570. DOI: 10.1016/j.carbpol.2009.09.007

[83] Oliveira IRWZ, Fatibello-Filho O, Fernandes SC, Vieira IC. Imobilização da lactase em micropartículas de quitosana obtidas por spray drying e usadas na construção de biossensores. Quimica Nova. 2009;**32**:1195-1201. DOI: 10.1590/S0100-40422009000500021

[84] Chávez-Servín JL, Castellote AI, López-Sabater M. Evolution of available lysine and lactose contents in supplemented microencapsulated fish oil infant formula powder during storage. International Journal of Food Science and Technology. 2008;**43**:1121-1128. DOI: 10.1111/j.1365-2621.2007.01588.x

[85] Yep YL, Li D, Mann NJ, Bode O, Sinclair AJ. Bread enriched with microencapsulated tuna oil increases plasma docosahexaenoic acid and total omega-3 fatty acids in humans. Asia Pacific Journal of Clinical Nutrition. 2002;**11**:285-291. DOI: 10.1046/j.1440-6047.2002.00309.x

[86] Serna-Saldivar SO, Zorrilla R, La Parra C, Stagnitti G, Abril R. Effect of DHA containing oils and powders on baking performance and quality of white pan bread. Plant Foods for Human Nutrition. 2006;**61**:121-129. DOI: 10.1007/s11130-006-0009-5

[87] Davidov-Pardo G, Roccia P, Salgado D, León AE, Pedroza-Islas R. Utilization of different wall materials to microencapsulate fish oil evaluation of its behavior in bread products. American Journal of Technology. 2008;**3**:384-393. DOI: 10.3923/ajft.2008.384.393

[88] Rebello FFP. Microencapsulação de ingredientes alimentícios. Revista Agrogeoambiental. 2009;**1**:134-144. DOI: 10.18406/2316-1817v1n32009223

Chapter 7

Lag, Constant and Decay Release Characteristic of St-PVOH Encapsulated Urea as a Function of Coating Thickness Using Different Empirical Models

Chigozie Francolins Uzoh and Stone Raphael Odera

Abstract

Uncoated urea, when applied to crops is susceptible to losses from volatilization, leaching, nitrous emission and water eutrophication. Plants require varying quantities of nutrients during different stages of growth; they need smaller amounts during infancy and larger amounts during the development of roots, stalk and stem. In this research, controlled release coated urea (CRCU) was synthesized via encapsulation with starch- polyvinyl alcohol biocomposite (St-PVOH). Lag, constant and decay release characteristic of the CRCU were simulated using different empirical models as a function of coating thickness. Structural elucidation and morphology of the raw urea and CRCU were determined using FTIR and SEM analytical techniques, respectively. FTIR confirm esterification reaction for St-PVOH. The SEM image of the raw urea appears rough and have fine openings while that of CRCU possesses a seemingly decrease in membrane porosity, ordered and uniform layer. This characteristic qualifies the CRCU as a semi-permeable membrane. Simulation results revealed that coating thickness of 4.3 and 6.4 are best desirable in designing a CRCU for plant at infancy stage, root, and stalk and stem development. Overall, sigmoidal law shows best robust prediction to expanded varying coating thicknesses.

Keywords: release kinetics, controlled release coated urea, empirical model, process simulation, diffusion

1. Introduction

Fertilizers play a vital role to supplement the nutrients required for plant growth [1]. However, plants require varying amounts of nutrients during different phases of growth. They need lesser amounts at the infancy and higher quantities at the development of roots, stalk and stem. Fertilizer demands therefore change intermittently throughout the plant growth. However, uncoated fertilizers are susceptible to losses from volatilization, leaching and fixation, and researchers have found that only about 30% of the fertilizers are used by plants while the rest is lost [2]. Such losses have adverse environmental impacts due to the addition of excess nutrients to air

IntechOpen

and water [1, 3]. Similarly, excess fertilizer spoils seedlings as only a small quantity of fertilizer is needed during sprout development [4–6].

Although coated urea provides a much longer release time and higher utility rate, it is mostly used in developed countries. It has not been popular in developing countries because of its higher cost [7]. It is noteworthy to say that developing countries, specifically Nigeria, consume more and more nitrogen fertilizer and yet have only 20–35% efficiency of nitrogen use.

Controlled release fertilizer (CRF) is a prominent green technology that helps to reduce the adverse effect of fertilizer on the environment. A controlled-release fertilizer (CRF) is a granulated fertilizer that releases nutrients gradually into the soil. The low solubility of the chemical compounds in the soil moisture determines the slowness of the release of the CRF. As conventional fertilizers dissolves in aqueous medium, the nutrients diffuse as quickly as the fertilizer dissolves. However, as controlled-release fertilizers are not readily water-soluble, their nutrients diffuse into the soil more slowly. The fertilizer granules may possess an insoluble substrate or a semi-permeable jacket that hindered the dissolution while allowing nutrients to flow out. Therefore, it is necessary to develop new types of fertilizers (CRF) that can improve nitrogen (N) use efficiency, sustain crop production, and protect the environment. Among newly developed commercial fertilizers, waterborne starch biopolymer-coated urea has great potential [8].

Starch is among the cheap, biodegradable, and abundantly available natural polysaccharide [9]. Starch alone is not realistic to be used as coating material due to its profound hydrophilic nature and poor mechanical properties [10]. Therefore, starch can be modified with some suitable additional agents to obtain coating materials devoid of these discrepancies. There is a plethora of studies for the use of starch as a coating material to produce controlled release devices [11, 12].

Coating film thickness plays a significant role for better controlled release properties [13]. The nutrient release time is a function of the diffusional path that the dissolved nutrient has to pass through and release from inside of the coating shell to the bulk water it is immersed in [14]. Critical review of literature reveals that viable controlled release properties of CRU are achieved mostly at the cost of non-biodegradability of coating material, process complexity, and elevated price [15].

Nutrient release from coated CRFs is generally controlled by diffusion through the coated layer [1, 16] and numerous recent studies have focused on predicting nutrient release behaviour. The simplest approach is regression reported and used in [17–21]. Kochba et al. [17] employed a semi-empirical model where the release of CRF was found to be a first-order process. However, the effects from geometry and size were ignored. Moreover, they failed to account for the lag period. Gandeza et al. [18] developed a semi-empirical model to investigate the effect of soil temperature on nutrient release from CNR-polyolefin-coated urea using a quadric equation: $CNR = \alpha + \beta(CT) + \gamma(CT^2)$. Wang et al. [22] also studied the effect of temperature on the release rate by using a regression model that reduced experimental time from days to hours. However, each of these models only related to a specific coating material. Recently, Azeem et al. [7] studied the effect of coating thickness on waterborne starch biopolymer coated urea, the study did not account for the lag, constant and decay release characteristics, expressly. In another account, Lubkowski [23] established that sigmoidal equation best correlate constant and decay release kinetics of chitosan coated fertilizer. A model developed by Than et al. [24] only simulates the second stage of nitrogen release from a coated urea.

From the foregoing, a clear understanding of lag, constant and decay characteristics of coated urea is still lacking. **Figure 1** illustrates the dissolution model of a spherical urea particle surrounded by water defined as the fluid zone. Urea particles comprise two parts: the urea core and the coating layer.

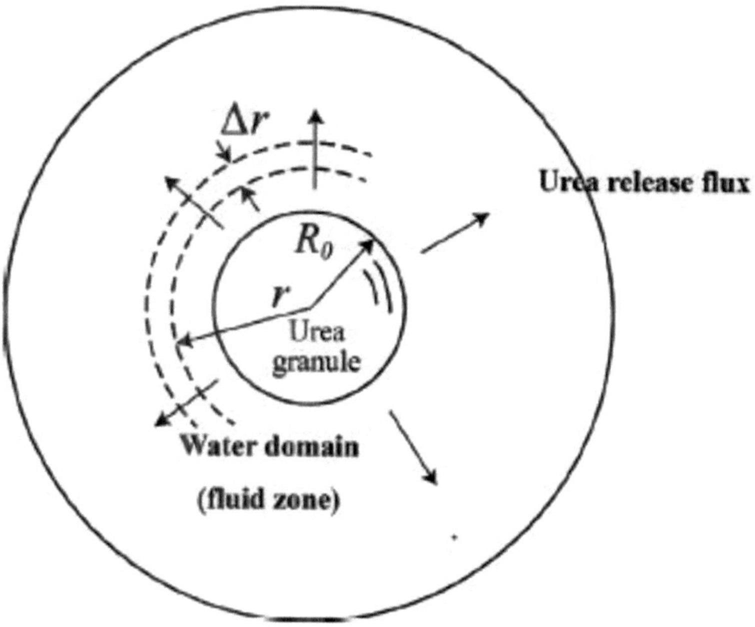

Figure 1.
Dissolution model of a spherical urea particle in water environment.

In this research, it is presumed that the coating layer was saturated with water at the time (t_0) of initial release i.e. lag period. Water inside the core initiates dissolving of the urea granules, where the concentration of the urea is kept constant at a saturated level provided that the solid urea is still inside the core. Nitrogen, from the urea, starts its release via the coating layer by diffusion at a constant rate and it is called "constant release" stage. When urea granule in the core is completely dissolved, urea concentration decreases, and a "decay release" stage starts and then proceeds to the end of the process. From on Raban's experiments, Shaviv et al. [16] suggested that the release profile from a single polymer coated granule had a sigmoidal shape comprising of these three stages. The cost of the coated controlled release fertilizer has made its usage limited and a paradigm shift has been observed on the frontiers to; use cost effective, environmentally friendly material to produce CRF and their nutrient release characteristics (lag, constant and decay). This research, therefore, covers the production of CRU using polyvinyl alcohol modified starch biopolymer as encapsulation material. The effect of coating thickness on the nitrogen release time and characteristics using different empirical models (power, exponential and sigmoidal law) were also investigated.

2. Materials and methods

2.1 Materials and pre-treatments

Fertilizer granules were procured from Eke Awka market in Awka, Anambra state. The granular fertilizer was subjected to sieve analysis to get a uniform range of the granules. For this study, granules of 2–2.8 mm were selected. The granules were carefully sealed to evade leaching on exposure to the atmosphere. Analytical grade polyvinyl Alcohol (PVOH) was obtained from CDH® and acetic acid were procured from JHD®. Refined starch was gotten from Burgoyne Burbidges & Co (India).

2.2 Preparation of starch-polyvinyl alcohol (St-PVOH) bio-polymer

Preparation of St-PVOH was done using two round bottom flasks. 10grams of polyvinyl alcohol was dissolved in 100 ml of deionized water at 90°C while aqueous solution of 10grams of starch dissolved in 100 ml of deionized water is added. A constant temperature of 90°C was maintained and stirred for 90 minutes to obtain a homogeneous hydrogel. The mixture was then allowed to cool to room temperature and an aqueous solution of acetic acid was added and stirred for another 90 minutes. A 25 grams by mass of starch was added to the resultant mixture to enhance its handling ability. The synthesized biopolymer was properly covered, kept at room temperature and consequently used for the coating of the granular urea.

2.3 Synthesis of St-PVOH encapsulated urea (St-PEU)

The St-PVOH coated urea was developed by creating balls of the synthesized biopolymer with the hands and fingers. Thereafter, the urea granule was encapsulated into each balls such that, the urea forms the core and the biopolymer, the sheath. These were repeated for five different coating thicknesses and then dry with the oven. A Constant diameter of the five different coating thicknesses was obtained by sieve analysis.

2.4 Determination of coating thickness of St-PEU granules

Vanier calipers were used to obtain the external diameter of the St-PVOH coated fertilizer. The average diameter of the urea granules were then subtracted from the measure diameter to obtain the thinking coating.

2.5 Determination of nutrient (nitrogen) release time from St-PEU granules

The dissolution of St-PEU in distilled water was studied by determining the time it took for the nitrogen present in the sample to be completely released (i.e. 100% release). As described in [7], two grams of each sample was immersed in a 200 ml of distilled water in a carefully sealed beaker. The period taken for the entire nitrogen released from the different coating thickness per time by using the Kjeldahl method was recorded. The experimental data related to release degree of mineral components from the starting fertilizer and from all prepared materials were described and interpreted with three kinetic equations: power, exponential and sigmoidal (Eqs. (1)–(3)):

$$\frac{M_t}{M_\infty} = k_p.t^n \tag{1}$$

$$\frac{M_t}{M_\infty} = 1 - exp\left(-k_e.t\right) \tag{2}$$

$$\frac{M_t}{M_\infty} = \frac{a - b}{1 + exp\left(\frac{t - d}{c}\right)} + b \tag{3}$$

where $Mt/M\infty$ is the fraction of nutrients released at time t, kp and n are the power equation constants, ke is the exponential equation constant, and a, b, c, d are the sigmoidal equation constants.

3. Results and discussion

3.1 Chemical formation of St-PVOH

Starch is hydrophilic in nature. The mechanical properties of starch are improved when it is blended with PVOH. Nevertheless, according to Izhar et al. [25], since both starch and PVOH are polar substances with hydroxyl group in their chemical structure, the blend of starch and PVOH exhibits poor water retaining properties. Therefore, acetic acid was introduced as a cross linker which enables the formation of intermolecular and intramolecular hydrogen bonds between hydroxyl groups of PVOH and starch to improve the integrity of St-PVOH blends. The chemical structure of St-PVOH blends can further be explained by the FTIR analysis as shown in **Figure 2**. The FTIR spectra of starch and PVOH were studied and compared with the spectra of the St-PVOH blend film. **Figure 2** shows the occurrence of -OH stretching vibration (3426 cm^{-1}), the intramolecular hydrogel (1646 cm^{-1}) and -CH_2OH stretching vibration (1260 cm^{-1}) as well as C-O-C ring vibration (928, 858 cm^{-1}) in the starch granules. For PVOH, a broad band at 3400–3100 cm^{-1} due to O-H stretching vibration and another band at 2936 cm^{-1} assigned as C-H stretching vibration were seen. The absorption points of 922 cm^{-1} denote C-C stretching. The 1200–1100 cm^{-1} region characterizes a number of modes, which have been shown to be sensitive to the degree of crystallinity in PVOH. The peak at 1147 cm^{-1} is crystallinity-dependent; this is in agreement with the work of Jayasekara et al. [26]. The FTIR spectra displays the following three sets of changes in the film: (a) the peak at 3300–3400 cm^{-1} of absorption bands weakened; (b) the crystallinity-dependent 1147 cm^{-1} peak of PVA weakened; and (c) the absorption peak of 922 cm^{-1} of PVA vanished and the peak of 854 cm^{-1} shifted.

The weakness, vanishing, and shifting of the characteristic absorption bands may be as a consequence of the interaction of different -OH groups in the starch and PVA molecular chains. It can be concluded therefore from these results that the starch was linked with PVA by chemical bonding introduced by acetic acid. This type of linkage has significant effect on the enhancement of compatibility.

3.2 Scanning electron microscope (SEM)

The coated urea was characterized using SEM to compare their morphologies to the original urea obtained. Morphology refers to the study of form and structure of a material and in this case, CRCU. SEM images were taken from the samples to present the morphology. As stated earlier, it must be noted that the starch is hydrophilic in nature due to the structure and composition. The SEM image of uncoated and coated urea (CRCU) is shown in **Figure 3** (a and b, respectively). The optical image for the uncoated urea is observed to appear rough and have fine openings where water can penetrate so as to dissolve the urea granule. An asymmetric structure and porosity are clearly seen at the magnification of ×1000. A seemingly decrease in membrane porosity, ordered and uniform layer, however, were noticed in **Figure 3**. This may be as a result of encapsulation with the biopolymer (St-PVOH). The slower the penetration time, the longer it will take for the urea to dissolve and consequently escape to the outer surface of the sample. This characteristic is needed for increasing the encapsulation efficacy of the released material, which in this research is urea. The elementary nanolayers of the bio-composite can be observed in the SEM image of **Figure 3a**. Using the image, it is difficult to determine the actual interlayer spacing of the biopolymer. In order to determine the d-spacing, high-resolution transmission electron microscopy would be needed. From **Figure 3b**,

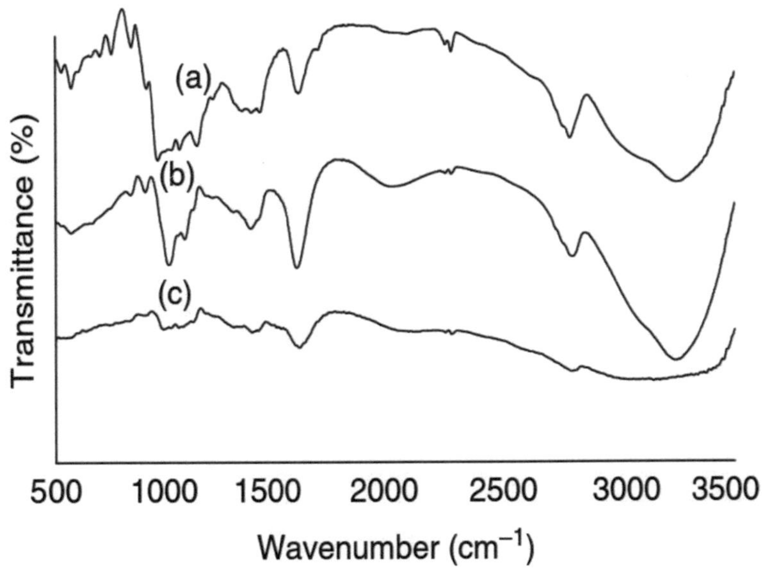

Figure 2.
FT-IR spectra of (a) PVOH, (b) starch, and (c) blend films (St-PVOH) cross-linked with acetic acid content.

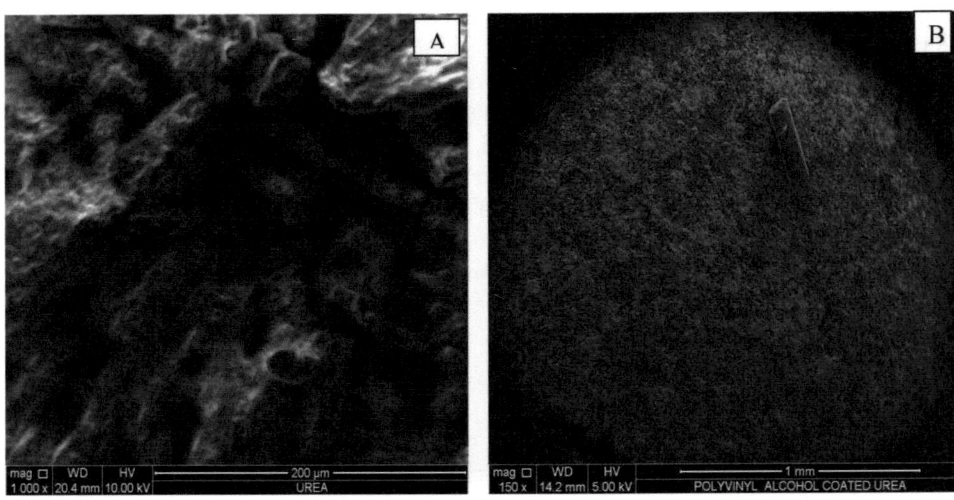

Figure 3.
SEM images of (a) uncoated urea (b) St-PVOH coated urea.

it can be concluded that the membrane can withstand the hydrostatic pressure of the dissolved nitrogen for long due to the tiny pore diameter. Thus St-PVOH biocomposite is an effective coating material for coated urea as it provides a semi permeable membrane and do not impede complete hindrance to nutrient release.

3.3 Effect of coating thickness on nutrient release time

The investigation of coating thickness is critical for the determination of nutrient release rate [13]. The result of the cumulative nitrogen release rate (wt%) for a monitored period of 30 days is given in **Figure 4**. **Figure 4** indicates that increase in thickness reduces the release rate. From **Figure 4**, the coated urea with 0.1 mm thickness released up to 19.82% nitrogen, that of 2.2 mm released 14.32% nitrogen, that of 4.3 mm thickness released 3.16% nitrogen, that of 6.4 mm thickness released

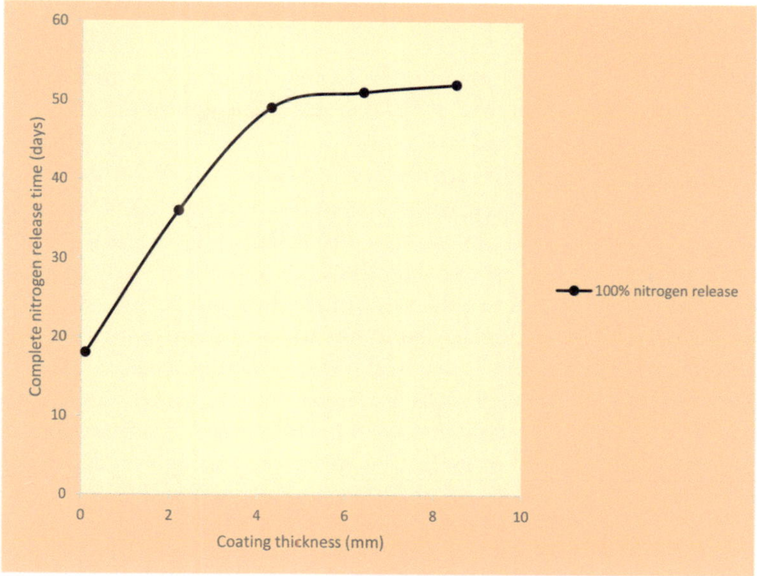

Figure 4.
Graph of complete nitrogen release time against coating thickness.

up to 2.93% nitrogen, and that of 8.5 mm thickness released 0.86% of nitrogen; all in 3 days. According to the Comité Européen de Normalization (CEN), a fertilizer can be described as having controlled release properties if the nutrient release is not more than 15% after 1 day or not more than 75% after 28 days [27]. From the above postulation, the coated urea with 2.2, 4.3, 6.4 and 8.5 mm can be considered controlled release fertilizer. A further look at the graph shows that the 8.5 mm coating gave the slowest controlled release rate followed by that of 6.4 mm.

The time required for each coating thickness to have released 100% nitrogen was also ascertained and the results are represented in **Figure 4**, which shows that the time required for 100% nitrogen release time increases with increase in coating thickness. Thus, the more the coating thickness, the slower the release rate and the longer the fertilizer will last on the farm. To further elucidate this phenomenon, it is important to understand the process of nutrient release from a polymer coated film. When a hydrophilic polymer is exposed to dissolution in water, the polymer matrix begins to swell due to uptake of water into the coating film [7]. Water moves through dynamic void between the macromolecules of the polymer film and dissolves the core nutrient as soon as it comes in contact with it [15]. This Osmotic uptake of water into the nutrient core is followed by the creation of hydrostatic pressure inside the coating shell. At a certain threshold value of the hydrostatic pressure, the swelled polymer matrix experiences deformation and the active nutrient starts to release from the microspores developed at the weakest sites on the coating film [28]. At that instant, the coating thickness plays an important role to control the release rate of the nutrient. Larger coating thickness results in higher release time as it takes longer time for the nutrient molecules to diffuse through thick coating film and reach the outer reservoir. Thus, the result obtained shows that the coating thickness of 8.5 mm have the highest release time and will last longer in the farm.

3.4 Release characteristic of the St-PVOH

As expected for a typical controlled release process, the nitrogen release of the CRCU generally decreases readily with increasing coating thickness, approaching

60% (for 8.5 mm thickness) in all the models indicating sufficient retardation of nitrogen. Such observations conform to a large extent with the results of literature [29]. In order elucidate this phenomenon, it is very important to understand the mechanism of nutrient release from a bio-composite or polymer coated film. When a hydrophilic polymer is subjected to dissolution in water, the polymer matrix begins to swell due to diffusion of water into the coating film [30]. Water moves through dynamic spaces between the macromolecules of the polymer film and dissolves the core nutrient as soon as it comes in contact with it [15]. This osmotic diffusion of water into the nutrient core is followed by the development of hydrostatic pressure inside the coating shell. At a threshold value of the hydrostatic pressure, the swelled network of the polymer experiences distortion and some active nutrient starts to release from the micropores developed at the weakest sites on the coating film [28]. At this point, coating thickness plays an important role to control the release rate of the nutrient. Larger coating thickness results in higher release time because it takes longer time for the nutrient molecules to diffuse through thick coating film and reach the outer reservoir [31]. In addition the augmented diffusion resistant for higher coating thickness also increases the release time [32]. Ko et al. [28], opined that a dire value of the coating thickness is essential for viable release properties. Nevertheless, coating thickness less than some critical threshold can cause immediate release of nutrient when subjected to dissolution test in water. This spontaneous release is due to the presence of void or microscopic pores in the coating film.

From the results of effect of coating thickness on release time as represented in **Figures 5–8**, it may be elucidated that when water comes in contact with the coated granule, it takes longer for water to completely wet thicker coating layer and reach the core nutrient. Once the core nutrient is dissolved, the swelled macromolecular hydrogel structure of encapsulated urea coating film bears the hydrostatic pressure of the dissolved nutrient for longer time in case of larger coating thickness. Once the dissolved nutrient reaches the interface, it again takes longer for its escape from thicker coating layer to the outer reservoir. Hence, thicker coating films result higher release time and better controlled release characteristics [24, 32]. This principle can only be attributed to those samples which resulted higher release time. On the contrary, there are samples which have very low release time. Shaviv et al. [16] reported that the release behavior of a single polymer coated granule had a sigmoidal shape comprising three stages: an initial stage during which no release is

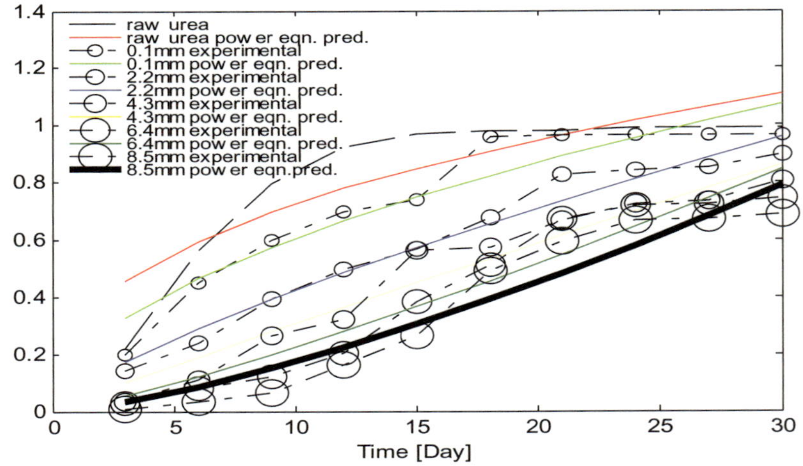

Figure 5.
Nitrogen release prediction of St-PVOH with power model.

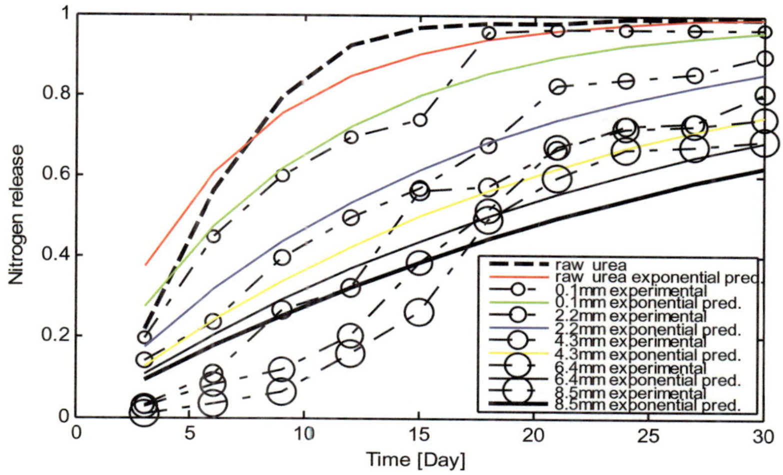

Figure 6.
Nitrogen release prediction with exponential model.

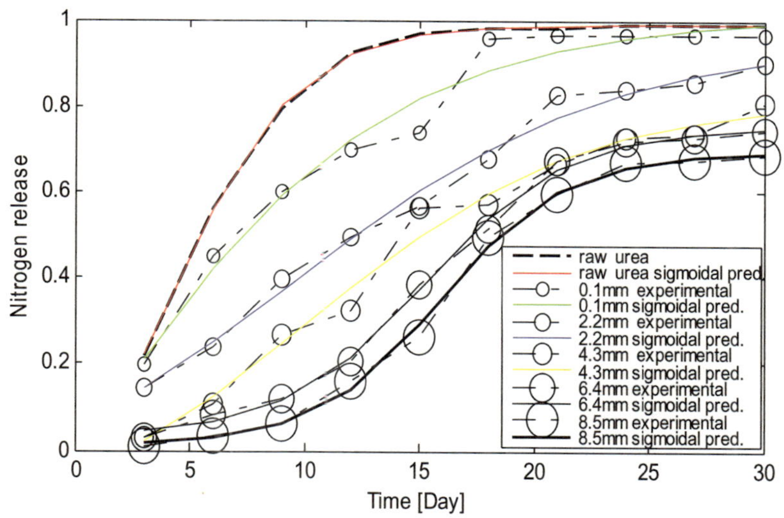

Figure 7.
Nitrogen release prediction with sigmoidal model.

seen (lag period); a stage of constant or steady state release; and finally, a stage of gradual decay of release. An interesting observation is recorded where the power and exponential model show intrinsic similarities in nitrogen release characteristics. One noticeable unusual result is the apparent lack of lag period for coating thicknesses of 0.1 and 2.2 mm. This type of nitrogen release profile cannot explicitly, support plants infancy stage. However, it is noticeable that coating thickness of 4.3 and 6.4 mm resulted in desirable release rate throughout the incubation time for all the models. The release degree of nitrogen from St-PVOH encapsulated urea were not more than 18% within 6ays incubation time and 62% after 30 days (see **Figures 5–8**).

A fertilizer is considered a controlled-release fertilizer (CRF) if these conditions are fulfilled: no more than 15% of nutrients are released within 24 hours, no more than 75% of nutrients are released within 28 days [27]. Specifically, a lag period of 2.5 days was observed with the sigmoidal model. The prediction ability of the sigmoidal model was further confirmed to be most robust from the observed

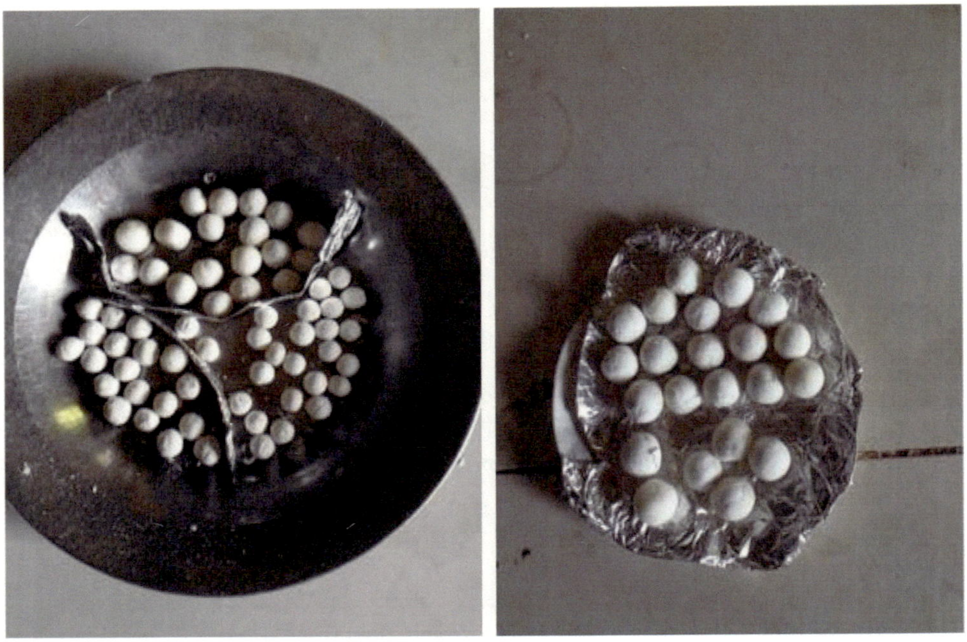

Figure 8.
Coated urea of various coating thickness.

	Power			Exponential		Sigmoidal				
	k_p	n	r	k_e	r	a	b	c	D	r
Raw Urea	0.2994	0.3847	0.888	0.1562	0.986	−0.3146	0.9900	2.7864	4.0190	1
0.1 mm	0.1838	0.5187	0.957	0.1066	0.987	−1.1582	1.0154	6.7805	−0.6027	0.988
2.2 mm	0.0769	0.7416	0.987	0.0638	0.991	−0.1634	0.9581	6.9069	9.7049	0.996
4.3 mm	0.0354	0.9381	0.976	0.0460	0.989	−0.1962	0.8166	5.8552	10.5014	0.993
6.4 mm	0.0143	1.1990	0.969	0.0385	0.972	0.0317	0.7519	3.1002	15.3357	0.999
8.5 mm	0.0075	1.3685	0.963	0.0325	0.965	0.0147	0.6912	2.6945	15.9864	0.999

Table 1.
Correlation coefficients of the empirical equations.

correlation coefficients established with the expanded varying coating thicknesses. Therefore knowledge of plant metabolic requirement is a prerequisite in the design of CRFs. The sigmoidal model has also proved its usefulness in the description of the controlled release. The constants and correlation coefficients of the sigmoidal model equations were determined and recorded in **Table 1**. *k*p and *n* are the power equation constants, *k*e is the exponential equation constant, and *a, b, c, d* are the sigmoidal equation constants.

4. Conclusion

The effect of coating thickness on lag, constant and decay release characteristic of CRCU has been investigated. It observed that the release time increases with increase in coating thickness. Starch-polyvinyl alcohol (St-PVOH) biocomposite as materials of biodegradable and regenerative resource was found to provide a semi permeable membrane suitable for the coating. Structural and morphology elucidation of the raw urea and encapsulated urea were established using FTIR and SEM instrumental analysis, respectively. There exist blend between the hydroxyl groups of starch and PVOH, where acetic acid served as a crosslinker to the bonding. Thus, the mechanical properties of the biocomposite are improved. SEM on the other hand, confirmed the porosity of the coating material, hence, validating its ability to absorb water. Observing the release rate, encapsulated urea with thicknesses of (2.2, 4.3, 6.4, and 8.5) mm is controlled release fertilizer, for they agree with Comité Européen de Normalization (CEN) descriptions of CRF. The result of the mathematical model shows that sigmoidal model satisfactorily fit the experimental data obtained. Conclusively, St-PVOH biocomposite is a feasible coating material in the production of controlled release fertilizer.

Acknowledgements

Tertiary Education Trust Fund (TETFUND), Nigeria, is hereby acknowledged for the research grant given to us to accomplish this study.

Author details

Chigozie Francolins Uzoh* and Stone Raphael Odera
Chemical Engineering Department, Faculty of Engineering, Nnamdi Azikiwe University, Awka, Nigeria

*Address all correspondence to: uzoh@unizik.edu.ng

References

[1] Plimmer JR, Gammon DW, Ragsdale NN. Encyclopedia of Agrochemicals. Vol. 1-3. 2003

[2] Al-Zahrani SM. Controlled-release of fertilizers: Modelling and simulation. International Journal of Engineering Science. 2014;**37**(10):1299-1307

[3] Oertli JJ. Controlled-release fertilizers. Fertilizer Research. 1980;**1**(2):103-123

[4] Shaviv A. Controlled release fertilizers, IFA International Workshop on Enhanced-Efficiency Fertilizers, Frankfurt. Paris, France: International Fertilizer Industry Association; 2005

[5] Trenkel ME. Controlled-Release and Stabilized Fertilizers in Agriculture. Paris: International Fertilizer Industry Association; 1997

[6] Trenkel ME. Slow- and Controlled-Release and Stabilized Fertilizers: An Option for Enhancing Nutrient Use Efficiency in Agriculture. IFA, International Fertilizer Industry Association; 2010

[7] Azeem B, KuShaari KZ, Zakaria B. Effect of coating thickness on release characteristics of controlled release urea produced in fluidized bed using waterborne starch biopolymer as coating material. Process Engineering. 2016;**148**:282-289

[8] Shoji S, Delgado J, Mosier A, Miura Y. Use of controlled release fertilizers and nitrification inhibitors to increase nitrogen use efficiency and to conserve air and water quality. Soil Science and Plant Analysis. 2001;**32**:1051-1070

[9] Guilherme MR, Aouada FA, Fajardo AR, Martins AF, Paulino AT, Davi MFT, et al. Superabsorbent hydrogels based on polysaccharides for application in agriculture as soil conditioner and nutrient carrier: A review. European Polymer Journal. 2015;**72**:365-385. 7-16

[10] Tang X, Alavi S. Recent advances in starch, polyvinyl alcohol based polymer blends, nanocomposites and their biodegradability. Carbohydrate Polymers. 2011:85

[11] Jin S, Wang Y, He J, Yang Y, Yu X, Yue G. Preparation and properties of a degradable interpenetrating polymer networks based on starch with water retention, amelioration of soil, and slow release of nitrogen and phosphorus fertilizer. Applied Polymer Science. 2013;**128**(1):407-415

[12] Niu Y, Li H. Controlled release of urea encapsulated by starch-g-poly (vinylacetate). Industrial and Engineering Chemistry Research. 2012;**51**(38):12173-12177

[13] Timilsena YP, Adhikari R, Casey P, Muster T, Gill H, Adhikari B. Enhanced efficiency fertilisers: A review of formulation and nutrient release. Journal of the Science of Food and Agriculture. 2015;**95**(6):1131-1142

[14] Ito R, Golman B, Shinohara K. Controlled release with coating layer of permeable particles. Journal of Controlled Release. 2003;**92**:361-368

[15] Azeem B, KuShaari KZ, Man ZB, Basit A, Thanh TH. Review on materials & methods to produce controlled release coated urea fertilizer. Journal of Controlled Release. 2014;**181**:11-21

[16] Shaviv A, Raban S, Zaidel E. Modeling controlled nutrient release from polymer coated fertilizers: Diffusion release from single granules. Environmental Science & Technology. 2003;**37**(10):2251-2256

[17] Kochba M, Gambash S, Avnimelech Y. Studies on slow release fertilizers: 1.

Effects of temperature, soil moisture, and water vapor pressure. Soil Science. 1990;**149**(6):339-343

[18] Gandeza A, Shoji S, Yamada I. Simulation of crop response to polyolefin-coated urea: Field dissolution. Soil Science Society of America Journal. 1991;**55**(5):1462-1467

[19] Medina LC, Sartain JB, Obreza TA. Estimation of release properties of slow-release fertilizer materials. Hort Technology. 2009;**19**(1):13-15

[20] Wilson ML, Rosen CJ, Moncrief JF. A comparison of techniques for determining nitrogen release from polymercoated urea in the field. Horticultural Science. 2009;**44**(2):492-494

[21] Zheng T, Liang Y, Ye S, He Z. Superabsorbent hydrogels as carriers for the controlled-release of urea: Experiments and a mathematical model describing the release rate. Biosystems Engineering. 2009;**102**(1):44-50

[22] Wang S, Alva A, Li Y, Zhang M. A rapid technique for prediction of nutrient release from polymer coated controlled release fertilizers. Open Journal of Soil Science. 2011;**1**(2):40-44

[23] Lubkowski K. Coating fertilizer granules with biodegradable materials for controlled fertilizer release. Environmental Engineering and Management Journal. 2014;**13**(10):2573-2581

[24] Trinh TH, Kushaari K, Shuib AS, Ismail L, Azeem B. Modelling the release of nitrogen from controlled release fertilizer: Constant and decay release. Biosystems Engineering. 2015;**130**:34-42

[25] Izhar SM, KuShaari K, Man Z, Samsudin YN. Influence of citric acid and curing time on water uptake. Applied Mechanics and Materials. 2014;**625**:123

[26] Jayasekara R, Harding I, Bowater I, Christie GB, Lonergan GT. Preparation, surface modification and characterisation of solution cast starch PVA blended films. Polymer Testing. 2004;**23**:17

[27] European Standard EN 13266. Slow Release Fertilizers – Determination of the Release of the Nutrients – Methods for Coated Fertilizers. Brussels: European Committee for Standardization; 2001

[28] Ko BS, Cho YS, Rhee HK. Controlled release of urea from rosin-coated fertilizer particles. Industrial and Engineering Chemistry Research. 1996;**35**:250

[29] Ding H, Zhang YS, Li WH, Zheng XZ, Wang MK, Tang LN, et al. Nutrients release from a novel gel-based slow/controlled release fertilizer. Applied and Environmental Soil Science. 2016:1-13

[30] Witono JR, Noordergraaf IW, Heeres HJ, Janssen LP. Water absorption, retention and the swelling characteristics of cassava starch grafted with polyacrylic acid. Carbohydrate Polymers. 2014;**103**:325

[31] Ito R, Golman B, Shinohara K. Controlled release with coating layer of permeable particles. Journal of Controlled Release. 2003;**92**:361

[32] Trinh TH, Shaari KZ, Basit A, Azeem B. Effect of particle size and coating thickness on the release of urea using multi-diffusion model. International Journal of Chemical Engineering. 2014;**5**:58